# Solid Particle Erosion
# and
# Erosion-Corrosion of Materials

## Alan Levy

**The Materials Information Society**

*Manager, Book Acquisitions*
Veronica Flint

*Production/Design*
Dawn R. Levicki

Library of Congress Cataloging Card Number: 94-73647

ISBN: 0-87170-519-2

SAN: 204-7586

**ASM International**®
Materials Park, OH 44073-0002

Printed in the United States of America

# ACKNOWLEDGEMENT

The author wishes to thank the Lawrence Berkeley Laboratory of the University of California for providing the facilities, personnel, and scientific environment that enabled him to pursue the intriguing subject of the solid particle erosion of materials with complete freedom and enthusiasitic support for seventeen years.

# DEDICATION

I dedicate this book to my wife, Bobi, who has had the faith and fortitude to share the life of a metallurgist for over forty years, when she didn't even know what a metallurgist was in the beginning.

# PREFACE

This book is intended particularly for the use of engineers—the *mechanical engineer* who designs and develops devices containing solid particle flows in benign and, especially, aggressive chemical atmospheres; the *chemical engineer* who develops and operates systems utilizing particle movement relative to containment surfaces in their operation; and the *materials engineer* who selects, predicts, and analyzes the performance of materials operating in erosive and erosive-corrosive environments.

To my knowledge, this is the first book to be devoted solely to the real-world occurrence of surface deterioration of materials by erosion and erosion-corrosion from a variety of carrier fluids and particles, in equipment used in the chemical processing, oil refining, and energy production industries. Real-world operating conditions were used in the laboratory to determine the behavior of the actual alloys and coatings used in industry to construct combustion boilers, reaction vessels, piping valves and a wide variety of other components. Descriptions of the mechanisms of surface loss are related to loss-rates under a variety of test conditions in a manner that affords the engineer the ability to make enlightened, empirical selections of alloys and process operating conditions.

Furthermore, this is also the first book to address in detail the combined erosion-corrosion mechanism and material losses that occur at elevated temperatures in the aggressive environments of fossil material combustion (in addition to coal) and process equipment. The behaviors described herein were obtained from controlled-condition laboratory tests of real engineering materials and operating conditions, offering the engineer data that is far more reliable than the data which can be obtained under the uncontrolled conditions occurring normally in the actual operation of the equipment.

From this information it should be possible to either reduce or eliminate uneconomical losses of material from components in systems whose surfaces are subjected to erosion or erosion-corrosion by small solid particles. These economical considerations include:

- initial cost of the materials of construction
- design of the components to minimize the most severe impact conditions—without unduly compromising the operation of the equipment
- manufacturing costs of the equipment
- selection of the operating conditions of the device—balancing component life with operating efficiency
- the time between overhauls
- inefficient operation of the device due to changes in the geometry of its components
- unplanned downtime due to part failure

Engineers and operators of equipment subject to erosion and erosion-corrosion material losses are given a basic understanding of the mechanism of surface loss of real engineering materials over a relatively wide range of representative industrial operating environments. The relationships of the mechanisms to material loss rates are described and supported by extensive test data in such a manner as to enable the selection of materials and operating conditions that should result in empirically predictable, economically acceptable loss rates and overhaul periods.

Clear reasons are given as to why it is not possible—at least at the present time—to use the analytical models of erosion in the literature to predict material loss rates under particular operating conditions to the level of accuracy required by industry.

Explanations are given for the physical and combined physical-chemical mechanisms by which material is removed from the surfaces of engineering materials exposed to the erosive forces of small, solid particles carried in gaseous or nonaqueous liquid flows. These mechanisms are related to the material loss rates of the metal alloys, ceramics, and coatings used in industry.

Using empirical techniques based on the established mechanisms of erosion and erosion-corrosion and simulation testing, a viable alternative to material loss rate prediction using analytical models is both proposed and explained. Also discussed is the selection of materials of construction, together with design and operating conditions that will result in economical solutions to problems caused by erosive material loss in industrial equipment.

Although not intended for the express use of either the scientist or the student, this book does provide valuable insights into the erosion process for all those interested in the subject.

*Alan Levy*
*April 1995*

# Table of Contents

## Chapter 1
## Introduction

## Chapter 2
## Mechanisms of Erosion

## Chapter 3
## Effects of Mechanical Properties of Metals on Erosion

# Chapter 4
# Effects of Erodent Particle Characteristics on the Erosion of Steel

# Chapter 5
# Erosion and Erosion-Corrosion of Steels at Elevated Temperatures

# Chapter 6
## Erosion-Corrosion of Materials in Elevated-Temperature Service

# Chapter 7
# Erosion of Protective Coatings

# Chapter 8
# Slurry Erosion

# Glossary of Abbreviations and Symbols

| | |
|---|---|
| **AFBC** | atmospheric fluidized bed combustor |
| **Al$_2$O$_3$** | alumina (aluminum oxide) |
| **bcc** | body-centered cubic |
| **BFBC** | bubbling fluidized bed combustor |
| **Bhn** | Brinell hardness number. Also abbreviated **HB** |
| **Btu** | British thermal unit |
| **CaCO$_3$** | limestone (calcium carbonate) |
| **CaO** | calcite (calcium oxide) |
| **CFBC** | circulating fluidizing bed combustor |
| **CGA** | coal gasification |
| **CrB** | chromium boride |
| **CVD** | chemical vapor deposition |
| **DBTT** | ductile-to-brittle transition temperature |
| **E-C** | erosion-corrosion |
| **EDS** | energy dispersive spectroscopy |
| $E_r$ | erosion rate |
| **FBC** | fluidized-bed combustor |
| **fcc** | face-centered cubic |
| **ft/s** | feet per second |
| **g** | gram |
| **gal** | gallon |
| **g/g** | gram/gram |
| **g/min** | gram per minute |
| **h** | hour |
| *h* | free-fall distance |
| **HB** | Brinell hardness number. Also abbreviated **Bhn** |

| | |
|---|---|
| **HRB** | Rockwell hardness number, B scale. Also abbreviated $R_B$ |
| **HRC** | Rockwell hardness number, C scale |
| **H₂S** | hydrogen sulfide |
| **HV** | Vickers hardness number |
| **IITRI** | Illinois Institute of Technology Research Institute |
| **JIT** | jet impingement test |
| **L** | liter |
| *L* | disk separation |
| **LBL** | Lawrence Berkeley Laboratory |
| **μm** | microns |
| **mg** | milligram |
| **min** | minute |
| **mp** | melting point |
| **m/s** | meters per second |
| **PCB** | pulverized coal boiler |
| **PFBC** | pressurized fluidized bed combustor |
| **P$_{O_2}$** | partial pressure oxygen |
| **PVD** | physical vapor deposition |
| *r* | barrel radius |
| *R* | radius to arc distance |
| *S* | arc length between erosion marks |
| **SEM** | scanning electron microscopy |
| **SiC** | silicon carbide |
| **SiO₂** | sand (silicon dioxide) |
| **SS** | stainless steel |
| *t* | time |
| *T* | temperature |
| *v* | angular velocity |
| *v*$_N$ | normal velocity component |
| *v*$_T$ | tangential velocity component |
| *V* | particle velocity |
| **WC** | tungsten carbide |
| **XRD** | x-ray diffraction |
| α | particle impact angle |
| θ | impact angle |
| ω | angular velocity |

# Chapter 1
# Introduction

The purpose of this book is to provide the reader with a comprehensive understanding of how and why small solid particles entrained in gas or liquid-fluid flows can cause degradation of the surfaces containing the flows by erosion and erosion-corrosion processes. This knowledge is essential in selecting the most economical materials of construction and operating conditions in the design of equipment as diverse as cement kilns, steam turbines, coal gasifiers, and combustion boilers. The book concentrates on particle velocities that are less than those that occur in gas turbines and higher-velocity equipment.

The word *erosion* derives from the Latin verb "rodene," which means to gnaw, or to wear away gradually. It is used in a diverse number of technologies, ranging from the geologist's loss of land or beach or mountain to the politician's loss of support to the engineer's loss of wall thickness. In all of these meanings, the process defies efforts to model it and thereby be able to predict accurately the amount of loss that will occur in a particular set of conditions. It is therefore important to understand the nature and mechanisms of erosive loss to effectively minimize its effects.

Erosion-material wastage is dependent on many interrelated factors that include the properties and structures of the target materials, the macroexposure and microexposure conditions, and the physical and chemical characteristics of the erodent particles. Combined erosion-corrosion at elevated temperatures is also a function of the chemical environment of the surface. The combination of all of these factors, sometimes exceeding 20 in number, results in material wastage rates that are peculiar to specific sets of conditions. Fortunately, erosion and erosion-corrosion reach steady-state rates in short time periods that can be extrapolated. So, while it is not possible to use rates of material loss obtained under one set of conditions for another exposure, it is relatively economical to obtain specific test data that is appropriate for each application.

Small solid particle impact erosion of materials occurs by the removal of material from a surface by a micromechanical deformation/fracture process. On ductile materials, the impacting particles cause severe, localized plastic strain to occur that eventually exceeds the strain to failure of the deformed material. On brittle materials, the force of the erodent particles causes cracking and chipping off of micro-size pieces. These two relatively simple mechanisms are the essence of small solid particle erosion. The particles are generally 5 mm (0.2 in.) in diameter or less, but can be much larger and move at velocities ranging from barely measurable to thousands of meters (feet) per second. In this book, velocities are limited to less than 300 m/s (900 ft/s), which are generally the limits of chemical processing or combustion operations. Aerospace erosion problems extend far beyond these modest velocities.

Solid particle erosion occurs on exposed components in many different types of equipment. It occurs by itself and in conjunction with other surface degrading mechanisms such as abrasion or aqueous or elevated-temperature corrosion. Some of the types of equipment with operations involving small solid particles are: heat exchangers in fluidized bed combustors and pulverized coal boilers, steam-turbine blades in electric-power generation plants, coal gasifier internals, and helicopter engine inlets and blades operating in sandy areas. Each of these erosion problems and others have been the subject of research to learn how to control the resulting metal wastage to tolerable levels.

The contents of this book are based upon some of the principal, refereed papers that the author has written over an 18-year period from 1975 on the subjects of the erosion and erosion-corrosion of metal alloys, their scales, and protective coatings for them in gas and liquid small solid particle erosive environments at room and elevated temperatures. Much of the information presented has retained the technical paper format from which it was taken.

## Types of Laboratory Testers

Three types of apparatuses have been widely used for solid particle erosion tests (Ref 1). The slinger system shown in its top view in Fig. 1.1(a) feeds erodent particles into the center of a rotating barrel in a chamber, whereupon they accelerate (sling) outward and exit from the barrel ends. Multiple samples can be arranged around the inner circumference of the chamber. To avoid hydrodynamic losses, the test chamber is normally held under vacuum using a roughing pump, but this has not been done in all cases. When the erodent particles exit the barrel, they will have a tangential velocity component $v_T$ and a normal velocity component $v_N$ due to centrifugal force. If ideal frictionless conditions are assumed for particle motion along the barrel with initial velocity zero at the barrel center, then

$$v_T = r\omega$$

$$v_N = \sqrt{2v_T}$$

where $\omega$ is angular velocity (rad/s) and $r$ is the barrel radius. An advantage of the slinger system is the fact that a large number of samples can be used during a test run. Impact angles are varied by tilting the samples. The disadvantages of the slinger system are its cost, operating complexity, the low effective dose per sample, and the large amount of particles required for each test. In general, the slinger apparatus is best suited for multiple samples to maximize erodent utilization. It is not used very often.

The rotating arm apparatus shown in its side view in Fig. 1.1(b) also is not used very often. Samples are mounted on the rotating arm (multiple-arm configurations can also be used). The erodent is introduced from above and allowed to free-fall a distance $h$ before being intercepted by the rotating samples. To avoid hydrodynamic losses, the test chamber is held under vacuum using a roughing pump. The erodent impact velocity is the tangential velocity $V = v_T = r\omega$ where

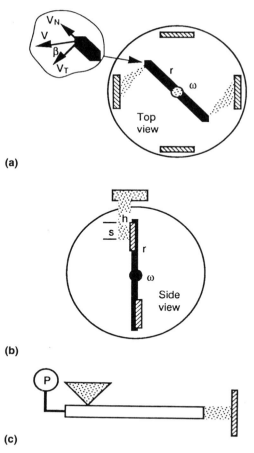

**(a)**

**(b)**

**(c)**

**Fig. 1.1**   (a) Slinger, (b) rotating arm, and (c) nozzle erosion testers

*r* is the arm radius and ω is the angular velocity. Impact angles can be varied by tilting the samples.

An advantage of the rotating-arm system is that control of the erodent velocity is simple, being fixed by the rotational speed of the arms (as also is the case for the slinger if β is held fixed). It also permits a wide velocity range. Compared to the slinger apparatus, it makes efficient use of the erodent because the entire dose is delivered to the target samples. The characteristics of the rotating-arm system are suited to low-erosion as well as high-velocity erosion test conditions.

The nozzle-tester shown in Fig. 1.1(c) is the most used type of erosion test equipment. It uses a jet of pressurized gas (air, nitrogen, inert gas) as a carrier to accelerate the erodent particles through the nozzle tube, whereupon they exit and impact a single target sample. Reactive gases and mixtures of gases can be used if the system is

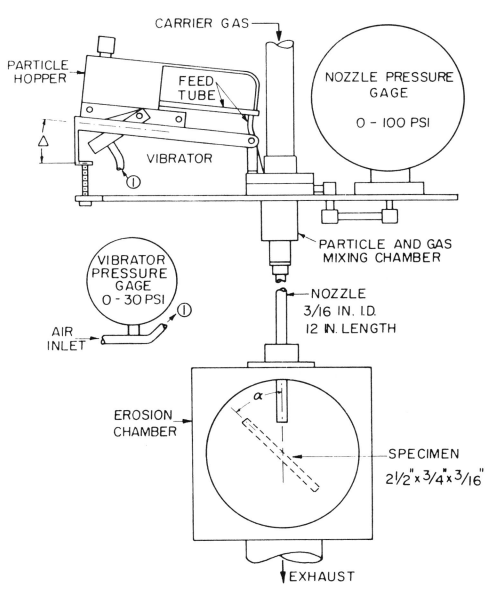

**Fig. 1.2**  Nozzle tester

contained in a sealed vessel. The tester can be oriented either horizontally, as shown, or vertically. The vertical flight tube orientation has been used for all of the room and elevated temperature testing discussed in this book. The erodent velocity can be varied by varying the air-pressure head, P. Erodent impact angles are varied by tilting the sample. Generally, the nozzle tube exit is located near enough to the target specimen surface to result in all of the erodent particles striking the target. The gas-stream system is relatively simple and inexpensive to construct and use compared to the rotating-type systems shown in Fig. 1.1(a) and (b). The nozzle tester apparatus does not require an enclosed test chamber (unless reactive fluids are used), dynamically balanced components, motor speed controls, or a vacuum pumping-filter system.

Furthermore, with a nozzle tester, the entire flux beam can be delivered to a single test sample

so that the amount of erodent used is known and is minimized for single-sample testing. The erodent velocity must be calibrated as a function of pressure head for each erodent used. An ASTM standard No. G76-83 (1983), exists for nozzle tester erosion systems. Temperature control from subzero to over 1000 °C (1832 °F) has been used in many investigations. For low particle impact velocities, the erodent particles can be gravity-dropped down a vertical nozzle tube, whose length determines the exit velocity.

The room temperature nozzle tester used to generate the ambient temperature data in this book is shown in Fig. 1.2. The velocity-measuring device is shown in Fig. 1.3. The elevated-temperature version of the same device is pictured at the beginning of Chapter 5. Both testers operate by feeding any dry, nonreactive, nonclumping, eroding particles from a vibrating hopper into a stream of gas. The erodent particles ranged in size

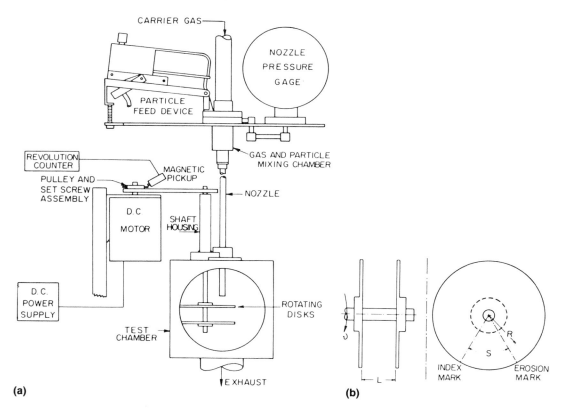

**Fig. 1.3**  (a) Rotating-disk velocity measuring device mounted in nozzle tester. (b) Variables for particle-velocity measurement: $L$, disk separation; $S$, arc length between erosion marks; $R$, radius to arc distance; $v$, angular velocity (rev/s); particle velocity $= V_p = 2\pi R v L/S$

from 20 to 1000 µm diameter. They were propelled down the nozzle tube at exit velocities up to 150 m/s for time periods from seconds to hours, impacting the specimen surface at angles of 15 to 90°. The pressure drop across the tube determines the impact velocity of the particles. The pressure is calibrated using the rotating disk device mounted in the tester, as shown in Fig. 1.3. The velocity is calculated using the computer program contained in Ref 2.

## Surface Degradation Research

The study of surface degradation by the impact of small solid particles has been carried out for many years. In earlier work, prior to about 1960 (Ref 3, 4), essentially all of the information published, in either book or technical paper form, consisted of compilations of material loss rate data of specific engineering materials obtained at particular sets of conditions experienced in specific pieces of operating equipment. There was essentially no effort reported to determine how and why the losses experienced occurred. Nor was any means presented to analytically predict material wastage rates under various combinations of operating or test conditions. Thus, the literature on erosion could only be used by those who had direct experience with the particular set of conditions reported. This situation resulted in only a small amount of information becoming generally available. Undoubtedly, a significant amount of additional erosion data was generated over the years for the private use of the companies that needed the data for their own product design or use.

Starting in 1958, with the publication of a proposed erosion model for ductile metals in a paper by Ian Finnie (Ref 5) in support of the development and use of catalytic crackers in the oil-refining industry, a plethora of papers on the erosion of metals by catalyst particles, ingested sand in helicopter engines, and several other applications were published (Ref 6-13). The majority of them followed the lead of Finnie and presented analytical prediction models accompanied by experimental data that was purported to demonstrate the predictive capabilities of the

models. Unfortunately, the models were not accurate enough in general to be used in engineering design, and the flurry of activity on erosion behavior ended in the early 1970s.

The primary weaknesses in the models were that they did not accurately predict metal-wastage rates without the use of empirically derived constants nor did they predict material losses over the whole range of the test conditions to which they were purported to be applicable. The models required considerable experimental test data to generate the values for anywhere from one to three constants in each of the several models that were proposed. The dependence on constants to be able to generate calculated data that was acceptably near experimental test data reduced the model equations to, essentially, statistically based empirical expressions that applied only to the materials and test conditions used to generate the data for each constant. Because the erosion process has not lent itself to analytical model representation, there is only a cursory review in this chapter of the information concerning the development and use of the models present in the literature.

Since the publication of models by Finnie, Bitter, Neilson and Gilchrist, and others, some additional effort has been reported (Ref 14) to use their models to build better, more predictive models without any more success than empirically developed constants could provide. Some of the more recent models contain up to eight different constants that require extensive test data to quantify.

A dividing point in the study of solid particle erosion occurred around 1975 when scanning electron microscopy (SEM) began to be used to observe the behavior of the eroded surfaces. The severe plastic deformation of eroded metallic surfaces produces surface textures with high and low elevations that are beyond the depth of field of optical microscopes and, therefore, are not amenable to observation by an optical microscope. The high magnifications that can be achieved using SEM along with the great depth of field scanning electron microscopes possess made possible the detailed observation of eroded surfaces that was necessary to gain an understanding of the mechanism of the erosion process on ductile metals. This capability was also of benefit in the determination of the mechanism of erosion of brittle materials. In Chapter 2, both of these mechanisms will be presented and discussed in detail. The beginning of the 1980s marked the advent of research based on microscopically observed behavior of the eroded surfaces rather than on speculation of what had occurred.

The use of SEM made it possible to observe what had physically occurred on eroded surfaces. Correlating this information with measured material wastage rates resulted in the establishment of surface degradation mechanisms that were more realistic, but they were more complicated and not as amenable to analytical modeling as the simple mechanisms such as microcutting that were speculated to have occurred in earlier work. The mechanisms consisted of sequential steps of microextrusion, forging and fracture for metals, and cracking and chipping for brittle materials, that led to eventual removal of material after a small number (~1 to 10) of particle impacts had occurred in a given microscopic area. Thus, the capabilities of the newer instruments that have been used to observe the erosion degradation of surfaces both increased the understanding of the erosion process and decreased the ability to accu-

rately represent the process by an analytical model. In the balance of this chapter, a brief review of the modeling work on solid particle erosion that has been in the technical literature prior to about 1980 will be presented. The primary purpose for including it here is to provide a level of continuity between the erosion research of the 1980s and 1990s and the earlier work prior to 1975.

The first effort of a scientific nature to study the erosion of ductile metals by small solid particles propelled against the target surface by a controlled blast of air through a tube (nozzle tester) was carried out by Finnie (Ref 5). He broke new ground in describing the erosion mechanism in a manner that was used to develop an analytical model of the erosion process. The model he developed was based on the premise that individual erodent particles traversed along the surface of the target metal, cutting out a swath of metal with dimensions that were those of the portion of the cross section of the particle that interfaced with the eroding metal and the length of the contact between the two (Fig. 1.4). The eroded material resisted this action by a property designated as its *dynamic flow pressure*. This work was carried out without the use of SEM and

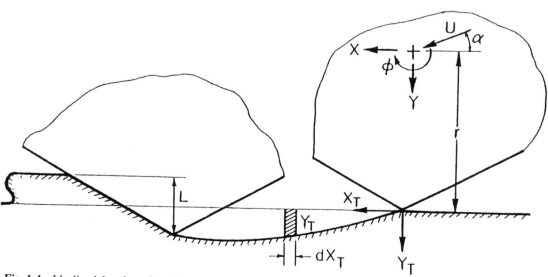

**Fig. 1.4**  Idealized drawing of erodent particle traversing a ductile metal surface

was based upon experimental observations of the surface scratches from the trajectories of individual particles of relatively large size that could be observed using optical microscopy.

Finnie used the equations of motion of the particles to analytically describe the occurrence of erosion. The volume of material removed was calculated from the trajectory of each particle and the amount of cutting that would occur as the result of pure plastic deformation with no cracks propagating ahead of the eroding particle. Important assumptions in his analysis were:

- The relation of depth of cut to the contact area on the particle
- The shape of the particle
- The distance along the surface of the metal that the particle and target were in contact
- How many particles out of the total that impacted the surface actually behaved in this manner

The resulting predicted erosion rates compared to experimental data are shown in Fig. 1.5.

Finnie's equations to represent the erosion of ductile metals were refined over the years, following the sequence shown in Table 1 (Eq T1, T2, T4, and T6). A final, further modified expression was published in 1978 (Ref 14) that involved moving the center of gravity of the microcutting particle deeper into the target metal surface than was used in his earlier work (see Fig. 1.4), where it would not be as subject to tumbling. Figure 1.5, Finnie's experimental (solid line) and predicted (dashed line) curves of the effect of impact angle on the erosion of a ductile metal, predicted a falloff of erosion to a zero loss at $\alpha = 90°$, which does not occur. The model also incorrectly predicted the impact angle of the peak metal wastage; it had to be arbitrarily moved in Fig. 1.5 to have the experimental and predicted curves coincide. It was inconsistent in establishing the effect of particle velocity (its exponent in the model). Another weakness of the model was having to select a low percentage of particles that behaved in the microcutting manner of all those that were directed at the target surface in order to obtain the amount of material wastage that was measured. It did, however, generally pre-

dict the shape of the material loss curve over the midrange of impact angles.

The continued modeling of the erosion process took several forms during the 1960s and early 1970s. Table 1 is a compilation of analytical models developed by several investigators that were basically inspired by Finnie's original effort. An excellent summary of this analytical effort is contained in Engel's book (Ref 15) in Chapters 4 and 5 and in a report by Majumdar et al. (Ref 16). Modifications were made to Finnie's basic thesis on the cutting mechanism to account for other observed types of plastic deformation and for aspects of the microcutting model for metals that were necessary to bring the resulting calculations more into line with measured erosion losses. The results did not significantly improve the ability of the models to predict erosion rates of metals because the basic mechanism assumption of many of the models was incorrect. The basic premise of most of the models for metal erosion in the literature was that the erosion process is a microcutting mechanism that can be analytically described by an equation of motion of the impacting erodent particle. The use of SEM to observe eroded surfaces has established that the erosion process in metals is primarily a sequential extrusion, forging and fracture phenomena and only rarely does microcutting occur on an eroding surface.

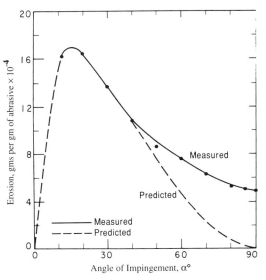

**Fig. 1.5** Experimental and predicted curves of erosion loss of aluminum

**Table 1.1    Analytical Expressions for Erosion of Ductile Metals**

**Eq T1 (Finnie 1960):**

$$Q = \frac{MV^2}{2P} [\sin 2\alpha - 3 \sin^2 \alpha]$$

$$\alpha \leq 18.5°$$

$$Q = \frac{MV^2}{24P} \cos^2 \alpha$$

$$\alpha \geq 18.5°$$

where $Q$ = volume of material, $M$ = mass of angular particles, $V$ = particle velocity, $P$ = plastic flow stress, and $\alpha$ = impingement angle.

**Eq T2 (Finnie 1961):**

$$Q = \frac{MV^2}{2p\left(1 + \frac{mr^2}{I}\right)} \frac{2}{P} [\sin 2\alpha \frac{2 \sin^2 \alpha}{P}]$$

where $M$ = mass of particle, $I$ = moment of inertia of particle, $r$ = radius of particle, $p$ = flow pressure, $P$ = K/[1 + $mr^2$/1], and K = particle constant.

**Eq T3 (Bitter 1963):**
Deformation erosion:

$$W_D = \frac{1}{2} \frac{M (V \sin \alpha - K)^2}{\varepsilon}$$

Cutting erosion:

$$W_{c_1} = \frac{2MC (V \sin \alpha - K)^2}{\sqrt{V \sin \alpha}} \left( V \cos \alpha - \frac{C(V \sin \alpha - K)^2}{\sqrt{V \sin \alpha}} \rho \right)$$

$$W_{c_s} = \frac{1}{2}M \frac{[V^2 \cos^2 \alpha - k_1(V \sin \alpha - K)^{3/2}]}{\rho}$$

where $W_D$ = erosion in units volume loss. $M$ = total mass of particles, $V$ = velocity of particles, $\alpha$ = impingement angle, K = constant, $k_1$ = constant from experimental data, $\varepsilon$ = energy to remove unit surface, $C$ = fraction of particles cutting ideally, $\rho$ = surface scratch term, $c_1$ = particle has final velocity, and $c_2$ = particle has no final velocity.

**Eq T4 (Finnie 1967):**

$$Q = C \frac{MV^2}{4} \frac{1}{P} f(\alpha)$$

where $Q$ = volume of material removed, $C$ = fraction of particles cutting ideally, $M$ = mass of angular particle, $V$ = particle velocity, $P$ = plastic flow stress, and $f(\alpha)$ = function of impact angle = 0.30 for $\alpha$ = 17°.

(Continued)

**Table 1.1   Analytical Expressions for Erosion of Ductile Metals   (Continued)**

**Eq T5 (Neilson and Gilchrist 1968):**

$$W = \frac{\frac{1}{2}M\,(V^2 \cos^2 \alpha - v_p^2)}{Q} + \frac{\frac{1}{2}M\,(V \sin \alpha - K)^2}{\varepsilon} \qquad \alpha < \alpha_0$$

**(a)**                                                      **(b)**

$$W = \frac{\frac{1}{2}MV^2 \cos^2 \alpha}{Q} + \frac{\frac{1}{2}M\,(V \sin \alpha - K)^2}{\varepsilon} \qquad \alpha > \alpha_0$$

**(c)**                          **(d)**

(a) is cutting erosion at small impact angles, (b) is deformation erosion at small impact angles, (c) is cutting erosion at large impact angles, and (d) is deformation erosion at large impact angles.

where $W$ = erosion, $M$ = weight of particles, $V$ = particle velocity, $\alpha$ = impact angle, K = velocity component normal to surface at erosion limit, $v_p$ = residual parallel component of velocity, $Q$ = unit of kinetic energy for cutting wear, $\varepsilon$ = unit of kinetic energy for deformation wear, and $\alpha_0$ = impact angle at which $v_p$ is O.

**Eq T6 (Finnie 1971):**

$$V = \frac{cMU^2}{4P\left(1 + \frac{mr^2}{I}\right)}\left[\cos^2 \alpha - \left(\frac{\dot{x}_o'}{U}\right)^2\right]$$

where $V$ = volume of material removed, $M$ = mass of eroding particles, $m$ = mass of individual particle, $I$ = moment of inertia of particle, $r$ = average particle radius, $U$ = particle velocity, $P$ = horizontal component of flow stress, $c$ = fraction of particles ideally cutting, $\dot{x}_o'$ = horizontal velocity of particle when cutting ceases, and $\alpha$ = impact angle.

**Eq T7 (Tilly 1972):**

$$\varepsilon_1 = \hat{\varepsilon}_1 \left(\frac{u}{u_r}\right)^2 \left[1 - \left(\frac{d_o}{d}\right)^{3/2} \frac{V_o}{V_r}\right]^2$$

$$\varepsilon_2 = \hat{\varepsilon}_2 \left(\frac{V}{V_r}\right)^2 F_{d,v}$$

where $\varepsilon_1$ = primary erosion, $V$ = particle velocity, $V_r$ = test particle velocity, $\hat{\varepsilon}_1$ = maximum primary erosion for $V_r$, $d_o$ = threshold particle size diameter, $d$ = particle size diameter, $V_o$ = threshold particle velocity, $\varepsilon$ = secondary erosion, $\hat{\varepsilon}_2$ = maximum secondary erosion for $V_r$, $V$ = particle velocity, and $F_{d,v}$ = fragmentation $d$ and $V$.

Ceramics and other low ductility materials erode by a cracking and chipping mechanism that is described in Chapter 2. It has also defied efforts to model it. Two representative models for brittle materials are:

- The model (Ref 17) that is based on the assumption that erosion occurs entirely by crack propagation and chipping
- The model (Ref 18) that is based on the assumption that plastic deformation contributes to the process of crack formation and surface chipping

When the basic erosion process is combined at elevated temperatures representative of those that occur in combustion or chemical reaction vessels with reactive gas or liquid carrier fluids and reactive particles, the physical-chemical processes that occur at the target surface modify the degradation mechanism and resulting material loss of the primary target material. The behavior that has been observed further directs the mechanisms away from those that can be depicted by an equation of motion expression or simple cracking and chipping behavior. Chemical as well as physical processes occur simultaneously. While they can be observed and their sequence and extent readily understood, they defy efforts to analytically model them.

An excellent recent paper (Ref 19) assessing the state of the art of analytical models for all of the wear processes as of 1993 has been written by Prof. Ken Ludema, a tribologist at the University of Michigan. In it, he painted a very bleak picture of the ability of the tribology field to develop a usable model for any wear process, that is, sliding, abrasion, and erosion. Ludema used solid particle erosion to illustrate the dilemma. He reviewed 98 mathematical erosion models gleaned from the literature. He discarded all but 28 of them because they were "not amenable to conclusive scrutiny." In the 28 remaining equations, he counted 33 variables being used by the authors. No model used more than 7 of them, and one model only used one, particle velocity, along with 2 constants. The number of constants used in the 28 models ranged from 1 to 8. There was no consistent pattern of use of the variables. Some used the same variable in the numerator that others used in the denominator. None of them could accurately predict material loss rates. Ludema commented on the sorry state that all of the wear processes were in with regard to the ability to analytically depict them in a consistent, practical, and usable manner for the design of products.

# References

1. Scattergood, R.O., Review of Erosion Test Methodologies, Paper No. 5, *Proc. Corrosion-Erosion-Wear of Materials at Elevated Temperatures* (Berkeley, CA), Jan 1990

2. Kleist, D.M., "One Dimensional, Two Phase Particulate Flow," M.S. thesis, Report LBL-6967, Lawrence Berkeley Laboratory, University of California, Berkeley, CA

3. Wahl, H., and Hartstein, F., *Strahlverschleias*, Franchlische Verlagshandlurg, Stuttgart, 1946

4. Wellinger, K. and Uetz, H., Gleit-, Spül-und Strahlverschleiss-Prüfung, *Wear*, Vol 1, 1957, p 225-231

5. Finnie, I., The Mechanism of Erosion of Ductile Metals, *Proc. 3rd U.S. Nat. Congr. Applied Mechanics*, American Society of Mechanical Engineers, New York, 1958, p 527-532

6. Finnie, I., Erosion of Surfaces by Solid Particles, *Wear*, Vol 3, 1960, p 87-103

7. Bitter, J.G.A., A Study of Erosion Phenomena, Parts I, II, *Wear*, Vol 6, 1963, p 5-21, 169-190

8. Neilson, J.H., and Gilchrist, A., Erosion by a Stream of Solid Particles, *Wear*, Vol 11, 1968, p 111-122

9. Raask, E., Tube Erosion by Ash Impaction, *Wear*, Vol 13, 1969, p 301-315

10. Tilly, G.P., Erosion Caused by Airborne Particles, *Wear*, Vol 14, 1969, p 63-79

11. Smeltzer, C.E.; Gulden, M.E.; and Compton, W.A., Mechanisms of Metal Removal by Impacting Dust Particles, *J. Basic Eng. (Trans. ASME)*, 1970, p 639-654

12. Head, W.J., and Harr, M.E., The Development of a Model to Predict the Erosion of Materials by Natural Contaminants, *Wear*, Vol 15, 1970, p 1-46

13. Tilly, G.P., A Two Stage Mechanism of Ductile Erosion, *Wear*, Vol 23, 1973, p 87-96

14. Finnie, I., and McFadden, D.H., On the Velocity Dependence of the Erosion of Ductile Metals by Solid Particles at Low Angles of Incidence, *Wear*, Vol 48, 1978, p 181-190

15. Engel, P., *Impact Wear of Materials*, Elsevier, 1976

16. Majumdar, S.; Natesan, K; and Sarajedini, A., "A Review of Solid Particle Erosion of Engineering Materials," Report No. ANL/FE-88-1, Argonne National Laboratory, Dec 1987

17. Sheldon, G.L., and Finnie, I., The Mechanism of Material Removal in the Erosive Cutting of Brittle Materials, *J. Eng. Ind. (Trans. ASME, Ser. B)*, Vol 88, 1966, p 393-400

18. Evans, A.G.; Gulden, M.E.; and Rosenblatt, M., Impact Damage in Brittle Materials in the Elastic-Plastic Response Regime, *Proc. R. Soc. (London) A*, Vol 361, 1978, p 343-365

19. Meng, H.S., and Ludema, K.C., "Wear Life Equations for Mechanical Designers: State of the Art," Keynote address, 1993 International Wear of Materials Conference (San Francisco), 1993

# Chapter 2
# Mechanisms of Erosion

## Erosion Mechanism of Ductile Metals

### Introduction

Structural metals have always had surface material removed in service as the result of erosion by small, solid, impacting particles. Through the years, particular engineering problems have arisen that temporarily intensified erosion research. In 1958, Finnie (Ref 1) developed an analytical model to attempt to predict erosion rates that was based on the assumption that the mechanism of erosion was microcutting, as described in Chapter 1. In spite of its shortcomings the microcutting mechanism was used by essentially all investigators in the field from 1958 to about 1975. Since then, however, a significant body of evidence has demonstrated that microcutting is not the primary mechanism by which ductile structural metals erode.

Another major consideration in the erosion of ductile metals that has been widely accepted for many years is the effect of the hardness and strength of the target material. A paper published in 1967 (Ref 1) contained a curve showing that the erosion resistance of annealed elemental metals increased with hardness from zinc up to tungsten. Based upon the behavior of the elemental metals, many people have used as an accepted fact, even a rule of thumb, that higher hardness results in greater erosion resistance. While greater hardness, in many cases, does result in

increased wear resistance in other types of wear, that is, sliding and abrasive wear, this basic premise does not apply to erosive wear, other than for the annealed, elemental metals of Ref 1. It is usually the case that increasing hardness has either no effect or a negative effect on erosion resistance.

This chapter is intended to present a body of evidence that describes the mechanism of erosion of metals based on physical observation of erosion surfaces at high magnifications with the great depth of field that is possible in scanning electron microscopy (SEM) correlated with measured metal losses (Ref 2). This mechanism of erosion can account for those aspects of erosion behavior that the microcutting mechanism could not, as well as those that it did explain. The mechanism consists of sequential plastic deformation processes that account for each of the separate occurrences that result in the overall surface degradation.

### The Platelet Mechanism of Erosion

A discussion of the evolution of the platelet mechanism of erosion of ductile metals by small solid particles (Ref 2) will be helpful in making the transition from the earlier accepted microcutting mechanism discussed in Chapter 1. In work conducted by the author to determine how specific steel microstructures affected erosion (Ref 3), an erosion weight loss curve depiction was used that resulted in basic doubt concerning the validity of the microcutting mechanism. In order to learn more about the initiation of the erosion that occurred, an incremental weight loss rate curve was used rather than the cumulative curve generally reported in the literature. The erosion was conducted on steel, incrementally, 60 g of particles at a time, and the weight loss caused by each 60 g increment was measured and plotted as shown in Fig. 2.1.

The initial erosion rate caused by the first 60 g of silicon carbide particles was much lower than that of subsequent 60 g batches of erodent. Also, extrapolating the curves down to 0 erosion shows that a number of grams of particles had impacted the surface before erosion losses began. With subsequent increments of impacting particles, the metal loss rate increased until it reached steady-state conditions where each increment of parti-

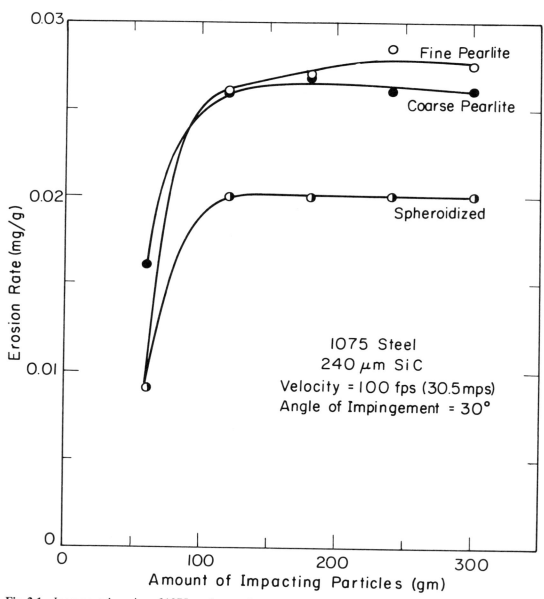

**Fig. 2.1**   Incremental erosion of 1075 steel to steady-state rate

cles caused the same metal-loss rate to occur. Steady-state conditions occurred after a relatively short exposure period.

If microcutting was the primary mechanism of erosion, the erosion rate of the initial, uneroded surface should be higher than subsequent incremental rates because work hardening of the surface due to the machining action would have reduced the machinability of the surface and the resulting amount of material loss. Weight loss by erosion also should have started with the first impinging particles that would have machined off metal. The curve in Fig. 2.1 also raised doubts concerning the effect of the hardness and strength of the metal on erosion. The lowest hardness, lowest strength, spheroidized condition of the 1075 eutectoid steel tested had the lowest erosion rate.

The effect of work hardening of the spheroidized steel was also investigated. It was expected that as the material was work hardened its erosion resistance would increase with the resulting hardness increase. Table 2.1 shows the erosion rate for the initial 60 g of erodent for 1075 steel specimens that were cold rolled to various percentage reductions prior to eroding them. The hardness doubled between the annealed steel and the 80% cold-reduced steel, but the initial erosion rate, rather than decreasing with increasing hardness, increased significantly. It did not achieve the steady-state incremental erosion rate, but approached it.

The two pieces of evidence presented in Fig. 2.1 and Table 2.1, coupled with micrographs showing extensive piling up of severely plastically deformed material that was extruded up and out of the craters produced by single particle impacts (Fig. 2.2), established doubts regarding the microcutting mechanism of erosion. These doubts were later reinforced when Fig. 2.3 was obtained. Figure 2.3 shows, at low magnification, the region at the edge of the primary erosion region of 1100-O aluminum that was eroded to the steady-state condition by spherical steel shot at a relatively steep impact angle of 60°. Throughout this book, impact angles in tables and figures are generally designated α. Many platelets can be seen that were made by first extruding microscopic bits of metal out of particle impact craters and subsequently forging them flat, much the same as occurs when a soft, malleable metal like gold is beaten with a ball-peen hammer. Some of the platelets are bent or cracked, indicating that another impact on them could break off the bent part or even the whole platelet. Some authors have used the term *ploughing* where *extrusion* is used in this book to describe the initial movement of metal on an eroding surface. Figure 2.2 shows that either term is appropriate.

## Microscopic Sequence of Erosion[Ref 4]

The development of an eroded surface on 1100-O aluminum was metallographically observed a few impacts at a time. This entailed developing a technique that would locate the same microscopic area using SEM after sequential, very short erosion exposures. To do this, microhardness indentation markers and much patience was required. Small erodent particles, typical of the sizes that actually occur in erosion environments, were used so that the type and extent of the damage caused by individual impacts would be representative of that which occurs in service. It was felt that the several millimeter diameter particles that have been used by others (Ref 5) and reported in the literature to observe the sequence of erosion forces produced

**Table 2.1   Effect of Cold Work on the Initial Erosion Rate of Spheroidized 1075 Steel**

| % cold rolled | Hardness (1000 gm load), HV | Initial erosion rate from 60 gm of SiC particles, gm/gm $\times$ 10$^{-4}$ |
|---|---|---|
| 0 | 152 | 0.98 |
| 20 | 242 | 1.03 |
| 40 | 262 | 1.49 |
| 60 | 288 | 1.66 |
| 80 | 316 | 1.72 |
| | | Steady-state erosion rate 2.2 |

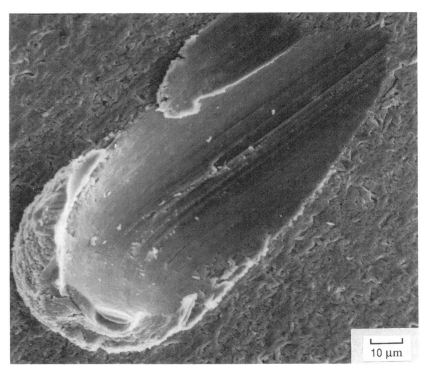

**Fig. 2.2**  Single-particle impact crater in 1100-O aluminum

damage and affected areas at the metal surface that were spread over so much greater an area than is produced by actual eroding particles that the resulting mechanism would not be representative of real conditions.

Figure 2.4 shows the appearance of the surface at relatively low magnifications at the beginning of the erosion exposure. Each 0.1 g of 600 μm silicon carbide particles consisted of 400 to 500 individual particles striking an area of approximately 1.25 cm in diameter. Only a few particles of the 500 struck the observed micro-area photographed in each incremental exposure, as can be seen by the limited amount of damage that occurred in the immediate area of the microhardness indentation markers.

Figure 2.5 shows a particular area of Fig. 2.4, within the marked box in Fig. 2.4(b) at a higher magnification. The changes in the pyramid-shaped area as it was struck by individual particles shows the development of flattened platelets as each subsequent 0.1 g of erodent struck the total eroding surface. It can be seen that after 0.4

g of erodent had been used, essentially no material loss had yet occurred in the observed area. Just the tip of the pyramid was knocked off, as shown in Fig. 2.5(d). This sequence accounts for the fact that there is a threshold period in erosion when severe plastic deformation is occurring with no material loss. It is followed by an increasing material loss as more particles strike the surface, until steady-state conditions occur (Fig. 2.1).

Carrying this sequential erosion experiment to steady state erosion produced the highly textured surface shown in Fig. 2.6. The magnified area in the figure is the region between the two sets of crossing parallel lines on the 7075 aluminum specimen pictured. It was eroded at an impact angle of 30°, which caused the elliptical shape. Hundreds of platelets in various states of deformation can be seen in the figure.

Figure 2.7 shows a single region of the eroded surface at higher magnification after the specimen shown in Fig. 2.6 was impacted with two additional 1 g increments of particles. The plate-

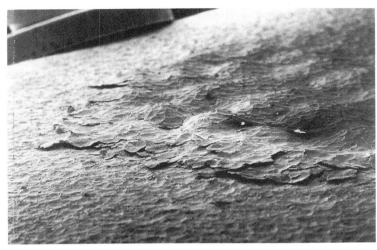

**Fig. 2.3** Platelets at edge of primary erosion zone on 1100-O aluminum. Velocity, 62 mps. α = 60. 20×

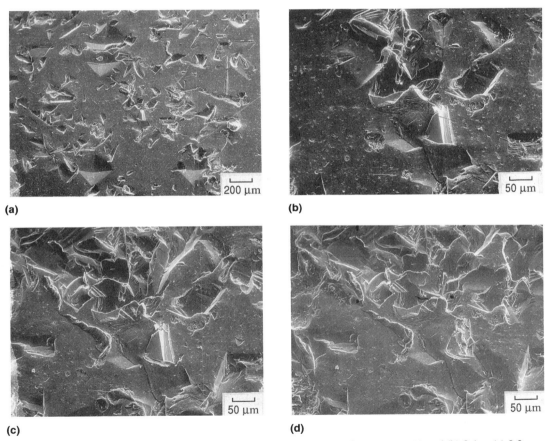

**(a)**

**(b)**

**(c)**

**(d)**

**Fig. 2.4** Appearance of surface of 1100-O aluminum early in the erosion process. (a) and (b) 0.1 g. (c) 0.3 g. (d) 0.4 g

lets present after the first gram are shown in Fig. 2.7(a). Figure 2.7(b) shows the marked changes that occurred after the second gram of particles struck the whole surface. The platelets in the upper right part of Fig. 2.7(a) have been knocked off, as has the platelet that sat astride the light line

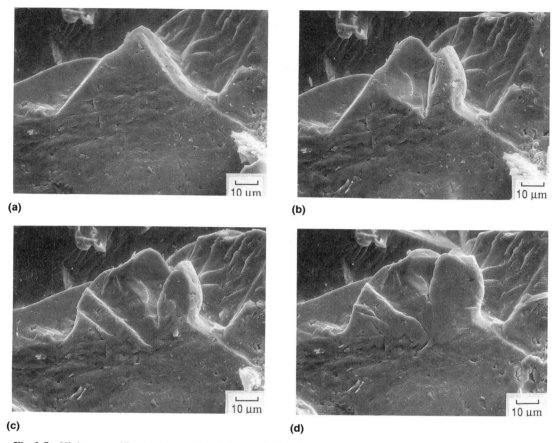

**Fig. 2.5** Higher magnification views of eroded area of 1100-O aluminum (from Fig. 2.4b). (a) 0.1 g. (b) 0.2 g. (c) 0.3 g. (d) 0.4 g

extending along the lower right side of both photos. New platelets have been extruded out from the main area of platelets in the center left side of Fig. 2.7(b).

These observations formed the basis for the development of the platelet mechanism of erosion. Many additional observations were made. A classic-shaped distressed platelet is shown in the center of Fig. 2.8 surrounded by other smaller platelets. It has the appearance of a thin pancake forging with considerable edge cracking and even some internal cracking. It is still attached to the substrate metal by a mushroom-like stem beneath it.

All of the evidence shown thus far was developed using aluminum alloys, which have an fcc crystallographic structure and many active slip systems. In order to determine whether the for-

mation of platelets is unique to metals of this type, several steel alloys were eroded and their surface microstructures observed. All of the alloys tested, 1020, 4340, and 304 stainless steel at various heat-treat conditions, formed platelets similar in size, shape and quantity to those observed on aluminum alloys. Figure 2.9 shows the eroded surface of 1020 plain carbon steel, which has a bcc crystallographic structure and comparatively few active slip systems, after erosion at an impact angle of 30°. It can be seen that the same mechanism of platelet formation occurred that was observed on the aluminum alloys. It also occurs at 90° impact, thereby accounting for the significant amount of erosion that was measured in 90° exposures (see Chapter 3).

In summary, the loss of metal from an eroding surface was observed to occur primarily by a combined extrusion-forging mechanism at all

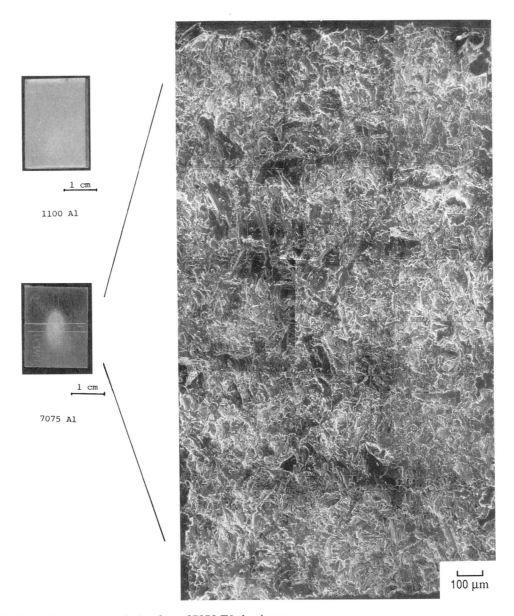

1 cm

1100 Al

1 cm

7075 Al

100 µm

**Fig. 2.6**   Steady-state eroded surface of 7075-T6 aluminum

particle-impact angles. The evidence indicates that the platelets are initially extruded from shallow craters made by particle impacts. Once formed, they are forged into the distressed condition shown in Fig. 2.8, in which condition they or parts of them are vulnerable to being knocked off the surface by subsequent particle impacts.

### Surface Observations

Evidence of extrusion being the initiating mechanism of platelet erosion was obtained in an experiment where a thin (3 µm) layer of copper was plated on a 1020 steel substrate that was subsequently eroded with a few silicon carbide

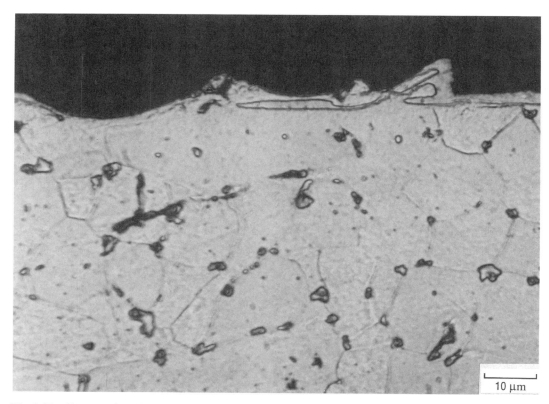

**Fig. 2.10**   Cross section of eroded surface of copper-plated steel

Figure 2.11 is a sketch of a proposed sequence of particle impacts that could cause the erosion shown in Fig. 2.10. The lip or platelet extruded out of the crater by the first impact is identical to one shown by Gulden and Kubarych (Ref 6) in their Fig. 10. Similar extruded lips are shown in several of the papers of Brown *et al.* (Ref 7, 8). The sequence of extrusion followed by forging of the extruded material can readily account for the surface and subsurface locations of the copper plating. The presence of the thin layer of copper over the entire surface of the craters in Fig. 2.10 indicates that it is extrusion that forms the lips or platelets rather than microcutting, as machining would have removed the thin copper layer.

Figure 2.12 is a sequence of micrographs of the surface of a 7075-T6 aluminum specimen at steady state erosion conditions. Figure 2.12(b) shows an extrusion-formed single platelet, Fig. 2.12(c) its subsequent spreading by forging and,

finally, in Fig. 2.12(d), its removal as the result of a particle striking it. The curved striations in a previously formed crater in Fig. 2.12(a) are covered over by the large platelet that was formed by one particle striking the lower center area shown in Fig. 2.12(b). The platelet was extruded from the resulting straight striation marked shallow crater and flipped to the left over the top of the crater formed earlier that has the curved striations in its surface. The striations on the surfaces of the craters are imprints of striations that form on the fracture surfaces of the silicon carbide particles used to erode the aluminum when they are crushed in a ball mill to obtain a desired particle size. Some of these striations can be seen on the sides of the silicon carbide particle shown in Fig. 2.13.

Figure 2.12(c) shows how two subsequent particle strikes in the area forged the platelet out to a larger size with a sub-platelet forming at its

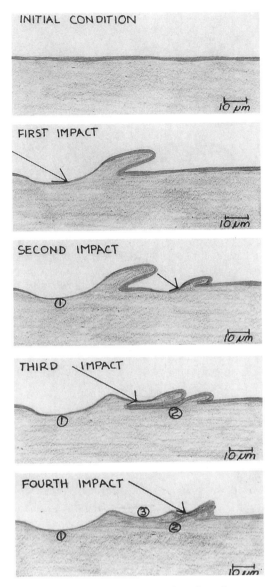

**Fig. 2.11** Proposed sequence of erosion of copper-plated steel

upper left side. In the fourth photo, Fig. 2.12(d), the platelet has been knocked off the surface and the curved striations in the earlier formed crater surface can be seen again (compare to Fig. 2.12a). A portion of the straight striations of the crater out of which the platelet was formed can also be seen again. Other common markings from step to step can also be compared to establish the action sequence.

The platelets do not adhere to the surfaces over which they are extruded and forged. Rather, they are attached to some location along the extrusion path. In the case of the platelet shown in Fig. 2.12(b), the platelet is attached at the point where its right side is next to the crater from which it was extruded. In Fig. 2.12(c), the attachment point is now beneath the forged-out platelet, making a mushroom-type configuration. The presence of platelets on eroding surfaces has been reported extensively in recent literature (Ref 6-10). The platelets have also been called chips, flakes, and lips.

### Surface Heating

While there is no direct evidence on any of the micrographs shown of heating of the eroding surface or any direct temperature measurements, considerable indirect evidence was gathered during the course of the investigation that temperatures near the recrystallization temperature of the target metal probably occurred at the immediate eroding surface. Figure 2.13 shows evidence of aluminum that has been melted and resolidified on the surface of a silicon carbide erodent particle that was used to erode 1100-O aluminum. Almost all of the eroding particles that were captured and observed after the erosion process had some once-melted aluminum on them.

Considerable recent work and some older work in the literature supports the fact that adiabatic shear heating and, possibly, some frictional heating occurs on the surface during the erosion process. Hutchings and Winter (Ref 11) discussed the generation of heat on an eroding surface. They attributed it to adiabatic shear heating when erosion lips were formed out of craters. Christman and Shewmon (Ref 12) saw evidence of melting of 7075-T6 aluminum when it was eroded with 5 mm steel balls that imparted very high forces and resulting deformation to localized erosion areas. Adiabatic shear bands were observed in the area where melting occurred. Shewmon (Ref 13) used the heating of the surface during erosion to calculate the effect of small particles. Gulden and Kubarych (Ref 6) saw evidence of melting on 1095 steel. Brown and Edington (Ref 14) used the low melting temperature metals gallium (mp = 29 °C) and indium (mp = 156 °C) specifically to show that erosion caused the

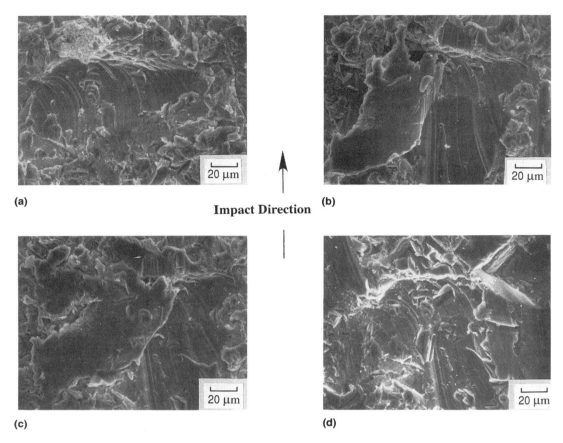

**Impact Direction**

**Fig. 2.12**   Sequence of platelet formation and removal on 7075-T6 aluminum

generation of sufficient heat on the eroding surface to melt both the gallium and indium in erosion experiments. Neilsen and Gilchrist measured temperature increases (Ref 15). Smeltzer et al. (Ref 16) also observed surface temperature increases. However, attempts to calculate the amount of heating that could occur on an eroding surface have not been able to account for the heating indicated by metallographic evidence (Ref 17).

### Work Hardening

The extensive plastic deformation that occurs on the surface of an eroding metal should cause some amount of work hardening to occur somewhere in the eroding surface region. Because the immediate surface was determined to be at an elevated temperature, it was thought that the work hardening might occur just beneath the heated surface. To determine where work hardening oc-

curred, microhardness measurements were made on cross sections of eroded aluminum and steel specimens. A very light, 5 g, load was used on the precision microhardness tester so as not to cause false readings near the surface. Figure 2.14 shows the microhardness survey of an eroded 1100-O aluminum specimen. The first measurements were made 5 μm beneath the actual surface. The lower hardness, immediate surface region can be seen, particularly for the 30° impact angle, where erosion rates are highest. The hardness increased to a subsurface work-hardened zone followed by a hardness decrease to that of the base metal. The evenness of the geometry of the microhardness indentations indicates that valid near-surface readings were obtained. In a paper by Salik and Buckley (Ref 18), a micrograph of an eroded 6061 aluminum cross section is shown that labels the three regions discussed above.

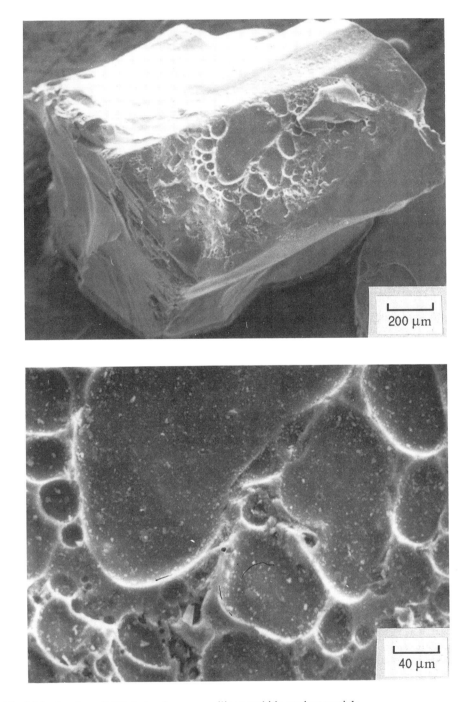

**Fig. 2.13**   Melted and resolidified aluminum on a silicon carbide erodent particle.

## *Platelet Mechanism Description*

Figure 2.15 is a sketch of a proposed cross section of an eroding ductile metal. The erosion-heated surface, 5 to 15 μm thick, consists of platelets at various stages of generation as the result of large plastic strain deformation. Beneath the platelet zone is a work-hardened zone that developed during the early stages of the erosion exposure. This zone lies beneath the heated surface region and strain hardens as a function of the strain-hardening coefficient of the target metal. Beneath the cold-worked zone is base metal at its initial condition.

Based upon the data presented earlier, it is proposed that the following sequence occurs in the erosion process. Initially, platelets are formed without loss of material (Fig. 2.4 and 2.5). Adiabatic shear heating of the immediate surface region begins to occur. Beneath the immediate

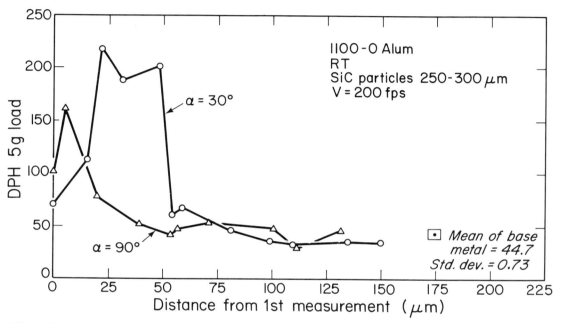

**Fig. 2.14**  Hardness survey of cross section of eroded area of 1100-O aluminum

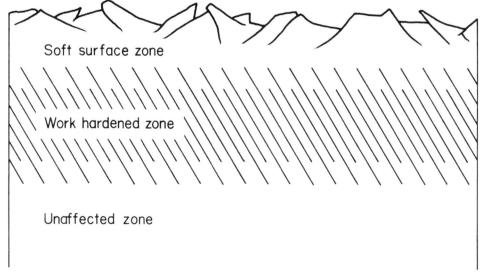

**Fig. 2.15**  Schematic of cross section of eroding metal surface

surface region, the mass of target material forms a work-hardened zone (Fig. 2.14) because the kinetic energy of the impacting particles is enough to result in considerably greater force being imparted to the metal than that required to generate platelets at the surface.

When the surface has been completely converted to platelets and craters (Fig. 2.6) and the work-hardened zone has reached its stable hardness and thickness, steady state erosion begins (Fig. 2.1). The higher steady-state erosion rate compared to the initial rates is due to the subsurface cold-worked zone acting as an anvil to increase the efficiency of the hammerlike impacting particles, to extrude/forge platelets in the now fully heated and most deformable surface region. When the anvil is fully in place and the platelets are fully formed and heated, maximum, steady-state, material removal rates will occur.

This cross section of material conditions will move down through the metal as erosion metal loss occurs.

### Erosion Mechanism of Brittle Scales on Metals[Ref 19, 20]

In most elevated-temperature erosion environments, the eroding surface is undergoing corrosion as well as erosion. The eroding surface region, therefore, is some combination of deposited erodent particles, surface scale, and base metal. To better understand the behavior of in situ formed, brittle surfaces on structural metal substrates undergoing combined erosion-corrosion, the erosion of scales formed under oxidation conditions on metal substrates has been studied. In one test series, a nickel oxide scale was formed up to 100 μm thick at 1000 °C on commercially pure nickel and subsequently erosion tested at 25

**Fig. 2.16** Cross section of duplex nickel oxide scale

°C using 250 µm silicon carbide particles at 30 and 100 m/s (Ref 19). In another series of tests 310 stainless steel was oxidized and combined oxidized-sulfidized in low partial pressure oxygen atmospheres at 980 °C, forming a thin $Cr_2O_3$ scale 2 to 5 µm thick, and then erosion tested at 25 °C using 50 µm silicon carbide particles at 60 m/s (Ref 20). While the $Cr_2O_3$ scale was much thinner than the NiO scale, both scales behaved in the same manner.

An analysis of the thick, duplex nickel oxide that formed on the nickel substrate is useful in gaining an understanding of what happens when erosive particles strike the brittle scale that forms on a ductile substrate. The cross section of the scale is shown in Fig. 2.16. The outer, darker-appearing layer is high density, large, columnar grains of nickel oxide. The inner, lighter-appearing layer consists of small equiaxed nickel oxide

grains with considerable porosity present, especially near the scale/metal interface. The outer scale was approximately 25 µm thick on all of the scales tested. The inner layer varied in thickness with the total thickness of the scale, being 75 µm in the case of 100 µm total thickness scale. The nature of the two scale morphologies is important in the erosion process.

Figure 2.17 shows the surface of the scale after a few particles have impacted. Note the tight grained surface the columnar grains presented to the impacting particles. Figure 2.18 shows a higher magnification micrograph of one of the plastically deformed craters that occurred in the brittle oxide. In the walls of the crater, radial cracks can be seen that emanate out from the impact crater. Figure 2.19 shows the condition that exists when the outer, more dense scale has been eroded away and the porous inner scale is

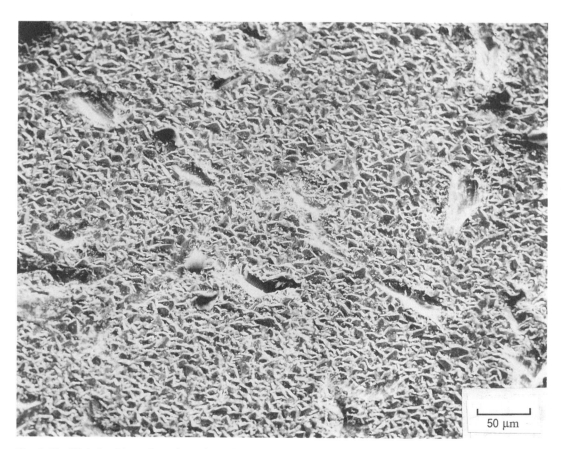

**Fig. 2.17** Nickel oxide scale surface after a few silicon carbide particle impacts

eroding. In Fig. 2.19(b), a large area of inner scale is exposed showing distinct Hertzian type cones that formed only in the porous inner scale. Figure 2.19(a) shows a location along the edge of the hole that is being eroded through the scale. In the photo, the outer columnar grain and the inner

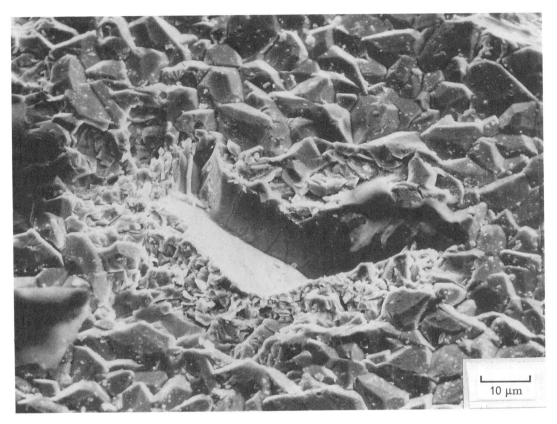

**Fig. 2.18**   Single-impact crater on nickel oxide surface

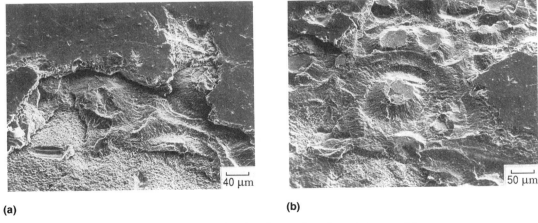

(a)                                             (b)

**Fig. 2.19**   Micrographs of eroding nickel oxide showing sections down to base metal

porous scale with some cone shapes showing can be seen. In the lower part of Fig. 2.19(a) substrate nickel with a crater formed by one impacting particle extruding up the beginning of a platelet at its right side can be seen.

Figure 2.20 shows sketches of the stress fields that develop inward from the plastically deformed crater produced by the initial particle striking the oxide surface. Planar cracks form along the interface between the two scale morphologies when the area is unloaded after the particle rebounds from the surface. This results in the cones having flat tops (Fig. 2.19). Thus, both radial and planar cracks form in the scale early in the erosion process, breaking up the surface region into a horizontal and vertical crack network.

During this time a threshold period occurs in the erosion process with very little, if any, scale being lost. Instead of the threshold period being used to make platelets and a subsurface work-hardened zone as occurs in metals, it essentially is used to crack the outer scale up into small

pieces that are separated from the inner scale by the planar cracks. Figure 2.21 is a sketch that shows that the Hertzian cones form under initial plastically deformed craters such as that shown in Fig. 2.18. All of this action serves to divide the scale into a mosaic of small, cracked areas that can be removed from the surface by subsequent particle impacts. The sequence is the same as that used to chip ice from a block of ice using an ice pick.

The removal of the scale is typical of a brittle material where the peak erosion occurs at 90° impact angle. Figure 2.22 is a plot of cumulative erosion weight loss versus the amount of impacting particles. A high rate of loss of nickel oxide scale occurred prior to the time that a bare nickel specimen (lowest curve) had lost any material. The threshold period for the nickel metal was of the order of 5 g. As the total thickness, h, of the nickel oxide scale increased, the amount of weight loss in the high erosion rate region of the curve increased before the straight line portion began. The straight line portion of each curve

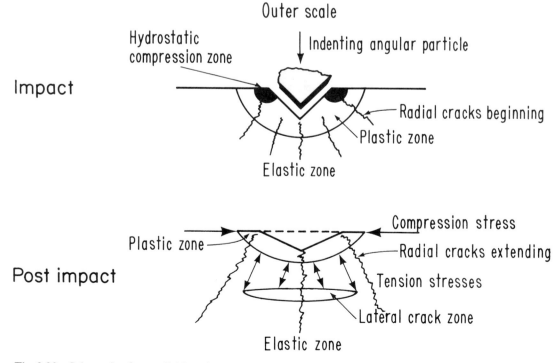

**Fig. 2.20** Schematic of stress fields and crack orientations under particle impact area

represents the period of the test where primarily nickel is being removed from the specimen along with some nickel oxide from the sides of the widening hole eroded through the scale layer.

Figure 2.23 is a plot of the incremental erosion rate experimentally measured for the first 0.5 g of impacting particles. It expands the left side of the curves in Fig. 2.22. The microstructures observed in Fig. 2.17 to 2.19 were also related to the erosion rates measured. A 90° impact angle where maximum erosion occurred was used in this test. It can be seen that the scale has a threshold period where the scale is being cracked, but none has been removed. At E1 for the 100 μm thick scale, weight loss begins to occur. At E2, the weight loss is primarily that of the dense outer scale layer that occurs at a lower rate than at E3.

At E3, the particles are eroding the porous inner scale layer and a maximum rate of erosion occurs. When the scale has been eroded off the nickel and the bare metal is undergoing its threshold period, the only weight loss is that of scale from the sides of the ever-widening hole through the scale. This loss occurs at an effective shallow angle and over a relatively small cross section and the erosion-rate curve, therefore, rapidly decreases, as shown.

The 20 μm thick scale specimen (lower curve, Fig. 2.23) had only the dense outer scale of nickel oxide. It reached its relatively low peak erosion rate early compared to the 100 μm thick duplex scale specimen and then fell off rapidly when only the sides of the hole were being eroded. At both times, the nickel substrate had not yet begun to erode.

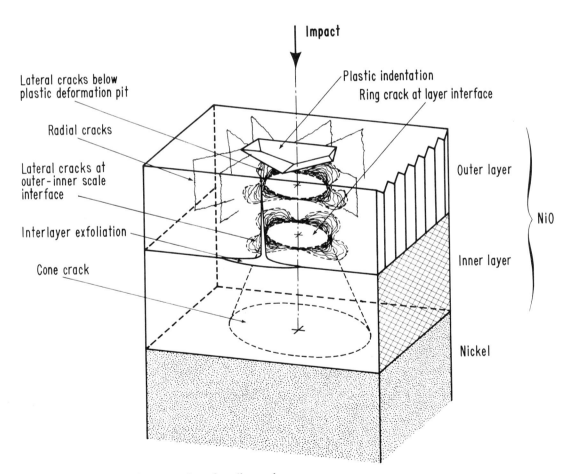

**Fig. 2.21**  Schematic of cross section of eroding scale

Figure 2.24 shows how the erosion rates of the nickel oxide compare at impact angles of 20° and 90° for two thicknesses of scale. The high peak rate at 90° occurs in the first 0.3 g of erodent, as was shown in Fig. 2.23. At the 20° impact angle, the dense, outer scale acts as an armor, protecting the porous, high erosion prone inner scale, and the rate of erosion is much lower in the early stages of the test. The peak erosion rate occurs at 5 g of particles, where the uncovered nickel first

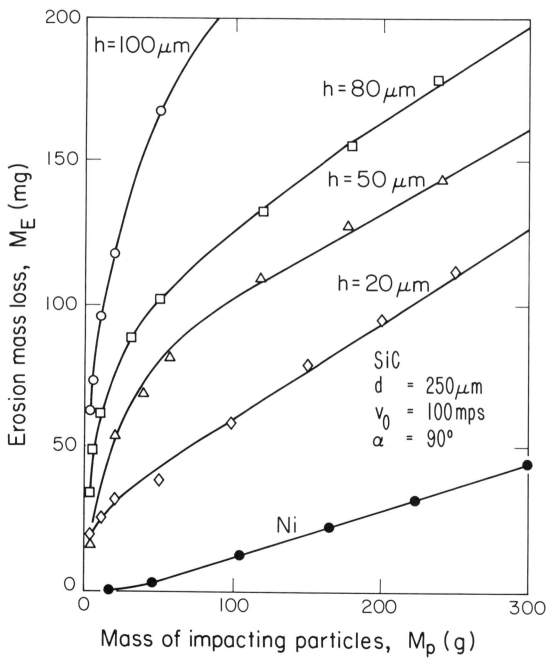

**Fig. 2.22**  Cumulative weight loss of nickel oxide scale versus amount of impacting particles

starts to undergo weight loss (bottom curve, Fig. 2.24). The lower erosion rate, at 20° compared to that at 90° impact angle, is typical of brittle materials.

From the investigation of the erosion behavior of relatively thick, duplex nickel oxide scale, it was determined that scales formed in situ on metals erode sequentially down through their thicknesses by a cracking and chipping mechanism rather than being knocked off the metal in pieces at the scale/metal interface. The presence of the ductile metal substrate does not appear to have a major effect on the way that the brittle scale erodes. However, the ductile substrate does have a small effect. The thinner scales can transfer more of the kinetic energy of the impacting particles to the ductile nickel and, hence, they crack and chip and then erode at lower rates than the thicker scales. As the thicker scales are removed, their rates of erosion decrease to a rate that becomes nearly the same for all scale thicknesses. In the case of the duplex nickel oxide scale, the harder, more dense, columnar grain outer scale protected the inner, softer, more porous, equiaxed scale from erosion as long as it was present.

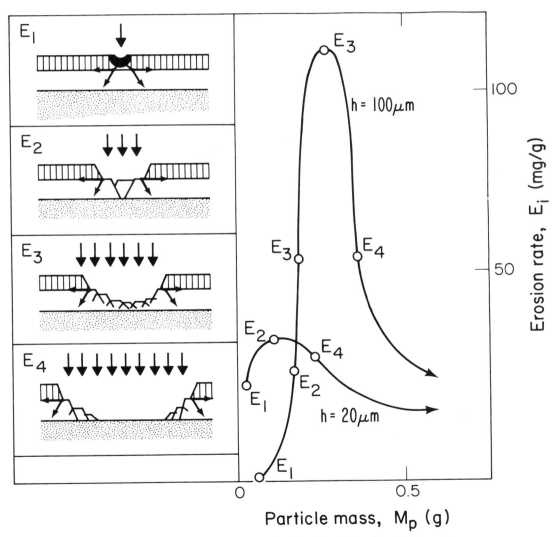

**Fig. 2.23**  Incremental erosion rate of nickel oxide scale during first 0.5 g of erodent impacts

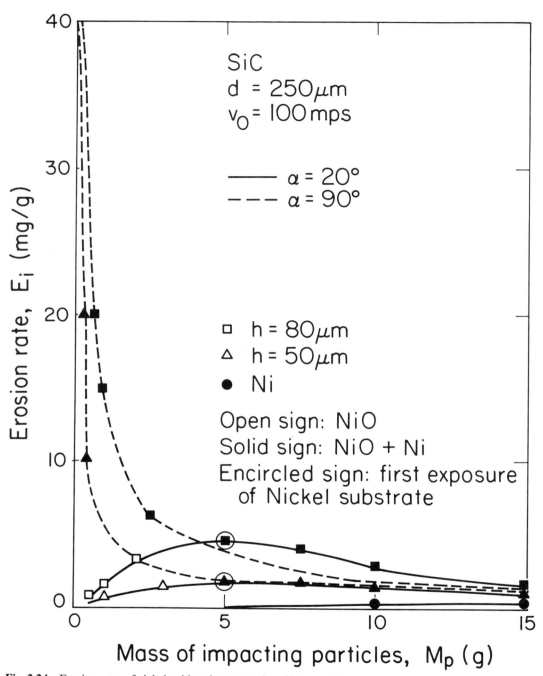

**Fig. 2.24**   Erosion rates of nickel oxide at impact angles of 20° and 90°

**References**

1.   Finnie, I.; Wolak, J.; and Kabil, Y., Erosion of Metals by Solid Particles, *J. Mater.*, Vol 2, 1967, p 682-700

2.   Levy, A.V., The Erosion of Metal Alloys and Their Scales, *Proc. NACE Conf. Corrosion-Erosion-Wear of Materials in Emerging Fossil Energy Systems* (Berkeley, CA), 1982, p 298-376

3. Levy, A.V., The Solid Particle Erosion Behavior of Steel as a Function of Microstructure, *Wear*, Vol 68 (No. 3), 1981, p 269-287

4. Bellman, R., Jr., and Levy, A.V., Erosion Mechanism in Ductile Metals, *Wear*, Vol 70 (No. 1), July 1981, p 1-28

5. Winter, R.E., and Hutchings, I.M., Solid Particle Erosion Studies Using Single Angular Particles, *Wear*, Vol 29, 1974, p 181-194

6. Gulden, M.E., and Kubarych, K.G., "Erosion Mechanisms of Metals," Report SR81-R-4526-02, Solar Turbines, Inc. San Diego, 30 Nov 1982

7. Brown, R.; Jun, E.; and Edington, J., Mechanisms of Solid Particle Erosive Wear for 90° Impact on Copper and Iron, *Wear*, Vol 74 (No. 1), Dec 1982, p 143-156

8. Brown, R., and Edington, J., Erosion of Copper Single Crystals under Conditions of 90° Impact, *Wear*, Vol 69 (No. 3), July 1981, p 369-382

9. Rickerby, D., and MacMillan, N., The Erosion of Aluminum by Solid Particle Impingement at Normal Incidence, *Wear*, Vol 60 (No. 2), May 1980, p 369-382

10. Christman, T., and Shewmon, P., Erosion of a Strong Aluminum Alloy, *Wear*, Vol 52 (No. 1), Jan 1979, p 57-70

11. Winter, R., and Hutchings, I., The Role of Adiabatic Shear in Solid Particle Erosion, *Wear*, Vol 34, 1975, p 141-148

12. Christman, T., and Shewmon, P., Adiabatic Shear Localization and Erosion of Strong Aluminum Alloys, *Wear*, Vol 54 (No. 1), May 1979, p 141-146

13. Shewmon, P., Particle Size Threshold in the Erosion of Metals, *Wear*, Vol 68 (No. 2), May 1981, p 253-258

14. Brown, R., and Edington, J., The Melting of Metal Targets during Erosion by Hard Particles, *Wear*, Vol 71 (No. 1), Sept 1981, p 113-118

15. Neilson, J.H., and Gilchrist, A., An Experimental Investigation into Aspects of Erosion in Rocket Motor Tail Nozzles, *Wear*, Vol 11, 1968, p 123-143

16. Smeltzer, C.E.; Gulden, M.E.; and Compton, W.A., Mechanisms of Metal Removal for Impacting Dust Particles, *J. Basic Eng.*, Sept 1970, p 639-654

17. Hutchings, I., and Levy, A., Thermal Effects in the Erosion of Ductile Metals, *Wear*, Vol 13 (No. 1), 1989, p 105-121

18. Salik, J., and Buckley, D., Effect of Mechanical Surface and Heat Treatments on Erosion Resistance, *Proc. Int. Conf. Wear of Materials* (San Francisco), 30 March-1 April 1981, American Society of Mechanical Engineers

19. Zambelli, G., and Levy, A., Particulate Erosion of NiO Scales, *Wear*, Vol 68 (No. 3), May 1981, p 305-332

20. Maasberg, J., and Levy, A., Erosion of Elevated Temperature Corrosion Scales on Metals, *Wear*, Vol 73 (No. 2), Nov 1981, p 355-370

# Chapter 3

# Effects of Mechanical Properties of Metals on Erosion

## Erosion of Steels and Aluminum Alloys(Ref 1)

The platelet mechanism of erosion changes many of the previously accepted relationships between erosion behavior and physical and mechanical properties of ductile metals. Ductility, strain hardening, "malleability" and thermal properties become more important, requiring that the effects of previously related properties such as hardness, toughness, and strength be reassessed. To do this, steel, aluminum, and nickel alloys (Ref 1) have been tested at various strength, hardness, and ductility levels using several compositions and thermal and mechanical treatments.

In Fig. 2.1 in Chapter 2, it was seen that a fine pearlite microstructure with a hardness of 99 HRB erodes approximately 40% faster at steady-state erosion conditions than does the same steel in the softer (79 HRB) more ductile, spheroidized condition. Figure 3.1 shows the incremental erosion rate curves for 1100-O aluminum and 7075-T6 aluminum. Both alloys formed the same type and size of platelets when eroded with silicon carbide (SiC) particles, but the erosion rates were markedly different. The 7075-T6 aluminum with a tensile strength of 76 ksi eroded 50% more than the much weaker 1100-O aluminum that has a tensile strength of 13 ksi. The 1100-O is much more ductile than the 7075 aluminum, 35% compared to 11% elongation. In the case of these

aluminum alloys, higher ductility results in lower erosion rates. Higher strength and hardness result in significantly greater erosion occurring.

Several steels were tested to determine the effects of properties on erosion behavior (Ref 2). The effect of ductility on the erosion rate of type 304 stainless steel is shown in Fig. 3.2. It can be seen that the less ductile, as-rolled steel has a higher erosion rate than the annealed steel. The effect of strain-hardening coefficient is shown in

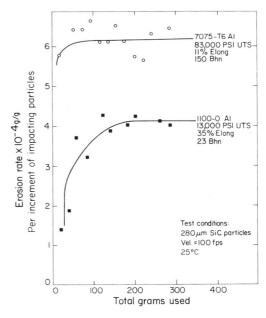

**Fig. 3.1**  Erosion rates of 7075-T6 and 1100-O aluminum

Fig. 3.3. The higher strain-hardening coefficient of the 304 stainless steel (0.55) compared to that of the 1020 steel (0.15) enables it to form its subsurface work-hardened zone considerably earlier than the 1020 steel can and, therefore, it reaches steady-state erosion in approximately 60 g of impinging particles while the plain carbon steel takes 120 g to reach steady-state erosion. In spite of the fact that the stainless steel forms a work-hardened subsurface layer sooner than the 1020 steel, it erodes at a much lower rate at steady-state conditions. While both alloys have similar strength properties, the 304 stainless steel has a 65% elongation while the 1020 steel has only a 25% elongation. Even though the strain

rates and actual deformation temperatures at the eroding surface are much higher than those of the slow strain rate tensile test that is used to determine elongation, tensile elongation has been able to be related to erosion behavior reasonably well in some steel alloys. However, in many other alloys, such as those described in the section "Erosion of Abrasion-Resistant and Heat-Resistant Alloys" in this chapter, ductility did not correlate with erosion behavior.

Another example of the effect of strain hardening of the subsurface "anvil" is shown in Fig. 3.4 and 3.5. In this experiment (Ref 3), the impacting particle was varied and the erosion of 1020 steel determined. A weak mineral particle that fragmented on impact, apatite, and a strong particle that did not, alumina ($Al_2O_3$), were used with all other testing conditions being the same.

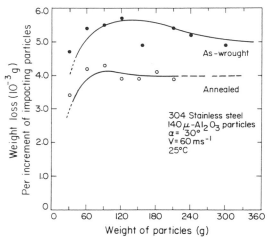

**Fig. 3.2**  Erosion rate of type 304 stainless steel

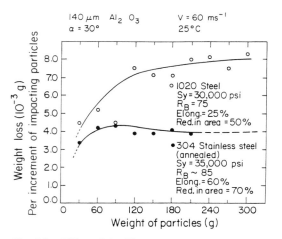

**Fig. 3.3**  Effect of ductility on the erosion of 1020 steel and 304 stainless steel

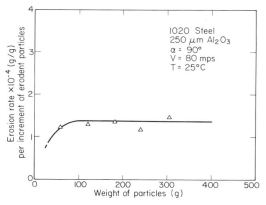

**Fig. 3.4**  Erosion rate of 1020 steel using $Al_2O_3$ erodent particles

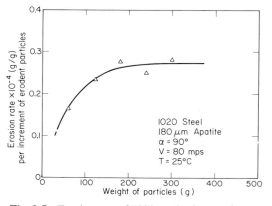

**Fig. 3.5**  Erosion rate of 1020 steel using apatite erodent particles

Figure 3.4 shows the incremental erosion curve using $Al_2O_3$ particles. The number of particles required to reach steady-state erosion was of the order of 50 g. Figure 3.5 shows the curve for the weak, apatite particle. Almost 200 g of particles were required to reach steady-state erosion even though the level of that steady-state erosion was only 25% as high as that which occurred when the $Al_2O_3$ particles were used.

It is postulated that the reason that the 1020 steel eroded with $Al_2O_3$ particles reached steady-state erosion in only one-fourth the particle flow it took for the apatite particle test is that the apatite fragmented into small particles when it impacted the surface. The effective size of the apatite particles after they broke up was too small with too little available kinetic energy to strain harden the subsurface layer as effectively as the $Al_2O_3$ did. It thus took longer to form an "anvil" in the 1020 steel, and it was not as effective as that formed by the $Al_2O_3$ particles. This effect of the breaking up of the apatite on the amount of kinetic energy that is available from its largest fragments may also relate to the particle-size effect in erosion. The largest fragments of the apatite were less than the 100 μm diam size below which erosion rate generally decreases with particle size. The effect of particle size is discussed in detail in Chapter 4.

One of the more interesting relationships between strength and hardness and erosion behavior is shown in Table 3.1 for 4340 low alloy steel (Ref 2). Four heat-treat conditions were used to determine the effect of property levels on erosion behavior. A change in ultimate tensile strength from 300 to 100 ksi and in hardness from 60 to 19 HRC had essentially no effect on the erosion resistance. If anything, the lowest strength and hardness condition, the spheroidized condition, had the lowest erosion. In this case, the elongation variation of three times did not have much effect on the erosion rate because even at the minimum 8% elongation in the as-quenched condition the 4340 was still ductile. It can also be seen in Table 3.1 that fracture toughness and Charpy impact strength had no effect on the erosion rates of this alloy.

Table 3.2 shows the effect of testing above and below the ductile-to-brittle transition temperature (DBTT) of 1020 steel (−18 °C) on its erosion resistance (Ref 2). The low-temperature test was run by strapping the specimen to a piece of dry ice (mp = −78 °C) before placing it in the erosion tester. It can be seen that the erosion rate goes up considerably when the steel is tested below its DBTT where it only has 1 to 5% elongation, compared to 25% elongation in its above-DBTT, more ductile condition.

There is considerable evidence in the literature to support the idea that higher strength and hardness do not result in greater erosion resistance within families of alloys. Gulden (Ref 4, 5) found that there was no difference in the erosion rates of 1095 steel specimens tested over a tensile strength range of 300%. These results compare

**Table 3.1   Effect of ductility, strength, toughness, and hardness on erosion behavior of 4340 steel**

| Heat-treat condition | Ultimate tensile strength, ksi | Hardness, HRC | Fracture toughness ($K_{Ic}$), ksi√in. | Elongation, % | Charpy impact strength, ft · lb | Steady-state erosion(a), mg |
|---|---|---|---|---|---|---|
| As-quenched | 307 | 60 | 34 | 8 | 10 | 1.03 |
| 200 °C temper | 273 | 53 | 58 | 11 | 16 | 0.97 |
| 500 °C temper | 182 | 39 | 62 | 14 | 12 | 0.97 |
| Spheroidize anneal | ≈100 | ≈19 | ... | ≈25 | ... | 0.90 |

(a) α = 30°, V = 30 m/s, T = 25 °C. Statistical average of weight loss per 30 g load of 140 μm $Al_2O_3$ particles at steady-state erosion

**Table 3.2   Effect of DBTT of 1020 steel on erosion**

| Test temperature, °C | Elongation, % | Steady-state erosion(a), mg |
|---|---|---|
| 25 | 25 | $0.25 \times 10^{-4}$ gm/gm |
| ~ −78 | 1-5 | $0.82 \times 10^{-4}$ gm/gm |

(a) α = 90°, V = 30 m/s, T = 25 °C. Statistical average of incremental weight loss per 30 g load of 140 μm $Al_2O_3$ at steady state

favorably with the 4340 test data in Table 3.1. Both test series were tested at a 30° impact angle. However, at a 90° impact angle Gulden found that the lowest strength and hardness heat treat conditions (30 HRC) resulted in a considerably lower erosion rate compared to the highest hardness (66 HRC) condition. This further substantiates the beneficial effect of greater ductility on erosion resistance.

In Ref 4, Gulden reported that 2024-T6 aluminum eroded considerably more than the weaker but more ductile 2024-O aluminum. This compares with the results shown in Fig. 3.1 comparing 7075-T6 with 1100-O aluminum. In further tests on 1095 steel, erosion rates were measured that were 2½ times higher for the 66 HRC, full hard condition compared to the 20 HRC, annealed condition. In Ref 6 and 7, Shewmon reported that higher hardness or fracture toughness did not enhance erosion resistance.

Work by Bahadur (Ref 8) to better understand the role of the mechanical properties of structural steels in their erosion behavior utilized precipitation-hardened, maraging steel in order to test over a relatively large range of properties and microstructures. A nozzle blast tester was used that propelled angular, 125 μm diam SiC particles at

the target at a velocity of 50 m/s and a feed rate of 20 g/min at an impact angle of 30° at 25 °C test temperature.

The aging of the solution-treated maraging steel was done at three different temperatures (480, 590, and 650 °C), for up to 24 h to obtain a complete spectrum of behavior in terms of the aged and overaged conditions. The effect of aging as indicated by the variation of hardness with aging time is shown in Fig. 3.6. Aging is accompanied by an initial rapid increase in hardness caused by increasing precipitation, which later gets moderated because of the start of austenite reversion. There is evidence of some precipitate particle growth occurring during the later part of aging as well. Austenite reversion increases the ductility of the steel. It should be noted that the aging temperature of 480 °C is commonly used in commercial practice.

The variation of erosion rate with aging time is shown in Fig. 3.7. Whereas aging for 1 min at the temperatures of 480, 590, and 650 °C results in the increase of Rockwell hardness by about 25%, 43%, and 48%, respectively, over that of the solution-annealed condition, the erosion rate remains practically unchanged. With further aging, the erosion rate continues increasing for the sam-

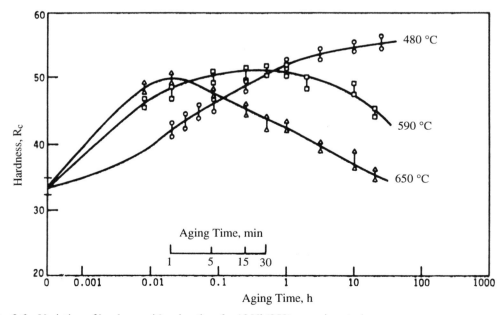

**Fig. 3.6** Variation of hardness with aging time for 18 Ni (250) maraging steel

ples aged at 480 °C, in which case no overaging was observed up to an aging time of 24 hours. Both the hardness and the erosion rate continued to increase. In the case of aging at 650 and 590 °C, whereas the peak hardness was reached in about 1 min and 30 min, respectively, the erosion

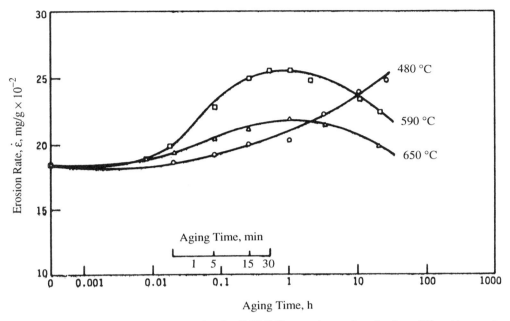

**Fig. 3.7**  Variation of erosion rate with aging time for 18 Ni (250) maraging steel aged at three different temperatures

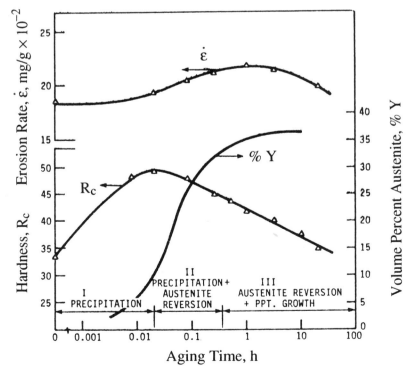

**Fig. 3.8**  Variation of erosion rate with hardness and austenite reversion for 18 Ni (250) maraging steel aged at 650 °C

peaked at 1 h and 30 min correspondingly. When aged further at these two temperatures, the hardness decreased and so did the erosion rate.

Continued aging in maraging steels results in the recovery of martensite, the formation of precipitates, and the reversion of nickel-rich austenite. The amount of austenite formed increases with aging time and temperature. It was determined that the erosion rate started to decrease for the 590 and 650 °C heat treatments when austenite reversion reached about 30 to 35%. The erosion rate continued to decrease after that because of softening due to further austenite reversion and precipitate particle growth. Figure 3.8 shows how the erosion rate, hardness, and austenite reversion, and precipitate growth interrelated for the 650 °C aging temperature.

The variation of erosion rate with ductility, while keeping the hardness essentially constant, was investigated by cold rolling the steel in the solution treated condition. The low-strain hardening coefficient of maraging steel, 0.025, makes this possible. Figure 3.9 shows the marked de-

crease in erosion rate with the increasing ductility indicated by the increase in the area reduction with decreasing percentages of cold work.

The erosion of 18 Ni (250) maraging steel is related to its aging behavior, which may be divided into three zones. In Zone I, hardness increases because of the precipitate formation, and the erosion rate also increases. In Zone II, along with precipitation, austenite reversion occurs and the net effect is softening of the matrix and leveling of the erosion rate. In Zone III, increased reversion of austenite and particle growth decrease hardness along with the erosion rate. The erosion rate varies directly with hardness when ductility remains unchanged during precipitation hardening. It was also found to vary inversely with the square of percent area reduction when the hardness remained constant. It shows that both higher ductility and lower hardness increase the erosion resistance of this material.

Thus, a body of evidence is available that relates higher erosion resistance to increased ductility in a number of alloys rather than to higher

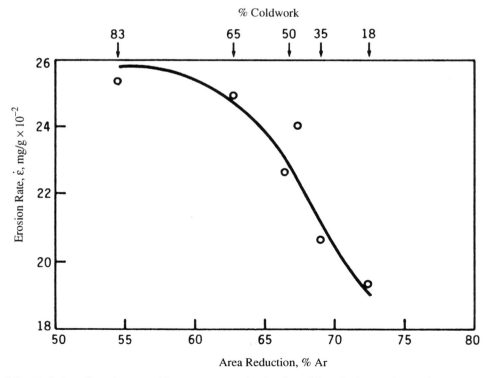

**Fig. 3.9**  Variation of erosion rate with percent area reduction for cold-worked maraging steel

strength and hardness. This behavior correlates well with the platelet mechanism of erosion of ductile metals. The ability to plastically deform to absorb the force from the kinetic energy of the impacting particles so that the local fracture stress of the metal platelets that are formed is not exceeded results in lower erosion rates. However, there is a limit to the effect that ductility has on increasing the erosion resistance of a ductile metal at the expense of strength. In Ref 9 that tradeoff

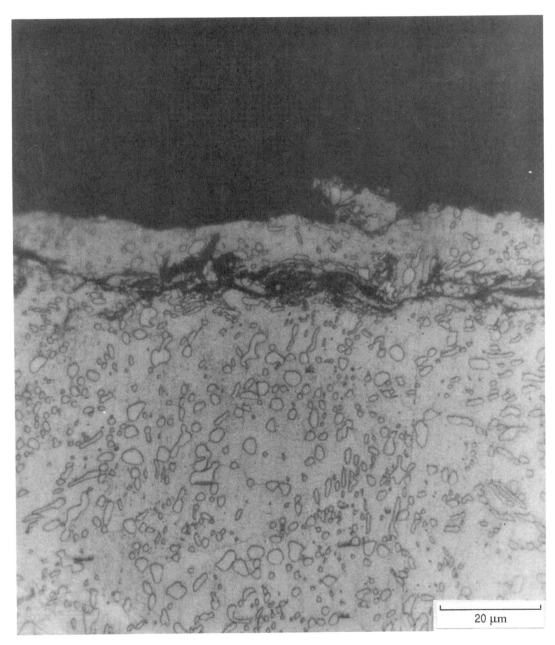

**Fig. 3.10**   Cross section of delaminated 1075 steel eroded specimen

was defined for 1020 steel. A point is reached where the strength of an alloy has been reduced to such a low level that localized fracture stresses can be exceeded and erosion rates begin to increase with further strength reductions, even though the ductility is still increasing.

Another mechanism of erosion, in addition to the platelet mechanism by extrusion forging, has been observed. It primarily occurs in multiphase alloys where at least one phase consists of isolated hard particles in a softer matrix of the major phase. This type of erosion produces relatively large, thick chunks compared to thin platelets that are knocked off of the surface by succeeding particles. It occurs in conjunction with the platelet mechanism of erosion. The mechanism is very similar to the delamination theory of wear developed by Suh (Ref 10) initially for sliding type wear. Jahanmir, who aided Suh in the development of the delamination wear theory, extended that concept to solid particle impact erosion (Ref 11).

The effort to adapt the sliding-wear mechanism to erosion was based on the structure of an eroded 1075 spheroidized steel cross section seen in Fig. 3.10. It shows a region approximately 7 to

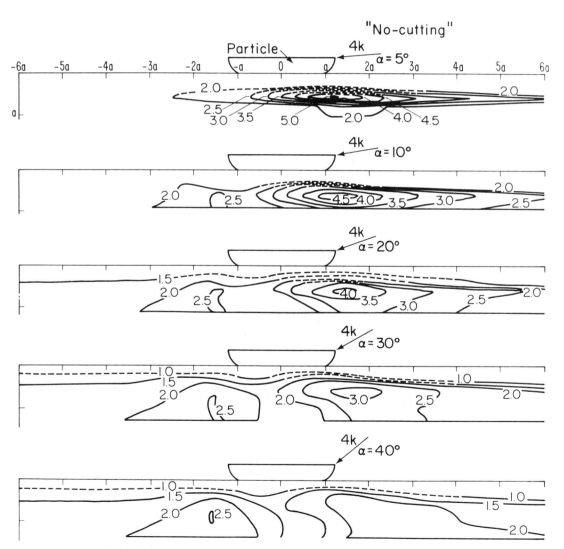

**Fig. 3.11** Plots of critical shear stress areas beneath an impacting particle

20 μm below the surface that was heavily voided and cracked. Its structure consists of small, hard, second-phase carbide particles in a soft ferrite matrix. The same type of subsurface voided and cracked zone is observed in sliding type wear; the difference is in the layer of metal above the delaminated zone. In sliding type wear this zone is a heavily deformed material, thought to be a dislocation sink area by Suh (Ref 10). As can be seen in Fig. 3.10, in erosion this area consists of essentially undeformed metal until the immediate surface is reached, where a surface platelet is seen. Branch cracks from the cracked zone periodically reach the surface, probably due to the surface forces of the extrusion-forging mechanism of platelet formation, and larger, thicker platelets are removed from the surface.

The stress-distribution model that was developed for sliding type wear was modified to describe the stress and strain distribution beneath an impacting erodent particle (Ref 11). Using the criteria for void formation in the original delamination theory, which is twice the shear yield stress, those regions under an impacting erosive particle where the shear stress equaled or exceeded the critical shear stress for void formation (with subsequent crack propagation from the void) were plotted. The critical stress is that which can exceed the cohesive bond between the hard, second-phase particle and the soft matrix.

The resulting plots as a function of the impact angle of the particle, 5 to 40° are shown in Fig. 3.11. If one combines the severity of the stress levels and the extent of the region under the impacting particle that exceeds 2 times the shear yield stress, it can be seen that the impact angle where both considerations peak is at about the 20° angle where maximum erosion is measured. Thus, there is some validity to the adaption of the delamination theory of wear to the erosion process.

There are several examples of subsurface void formation and cracking in the literature. In Ref 4, Gulden and Kubarych observed it when erosion testing 2024 aluminum. Brown and Edington observed it in erosion tests of copper and iron in Ref 12 and 13.

## Erosion of Abrasion-Resistant and Heat-Resistant Alloys[Ref 14]

The erosive wear behavior of a number of alloys commonly used for their abrasive wear or elevated temperature strength properties were studied to determine how variations in their mechanical properties by thermal or working treatments would affect their performance. Alloys were selected that had particularly large variations in such properties as hardness and strength between their annealed and treated conditions. The results of the investigation clarified the role of those properties that had been used as criteria for the selection of materials for all kinds of wear service, including erosive wear. Table 3.3 lists the alloys and the basis for their selection in the study.

Table 3.4 gives the steady-state volume erosion rates of the alloys tested, using silicon carbide and quartz erodents. Elongation and Vickers hardness values for the alloys are also included in this table. For the group of wrought, heat-resistant alloys (Haynes 188, Hastelloy C-276, and Cabot 718), differences in elongation from 9 to 67% and in hardness from 229 to 476 HV had essentially no effect on any of the erosion rates measured. In the case of silicon carbide erodent, the strengthening treatments of the target alloys made negligibly small differences on the erosion rates. For quartz erodent, the differences were somewhat larger, but still rather small. It is clear from Table 3.4 that not only were erosion rates higher when silicon carbide erodent was used, but that the response with respect to impact angle was closer to that expected for ductile materials than when quartz erodent was used. For silicon carbide erodent, the erosion rate at a 90° impact angle was always less than it was at 30° impact, although the difference in rates between these angles was small. The results of tests employing quartz erodent showed a trend of maximum erosion occurring at a higher impact angle than for silicon carbide. With the exception of Ferralium 255, which showed a continuous decrease in erosion rate with impact angle for the limited number of angles tested, and Haynes 6B, which showed a continuous increase, all of the alloys had a peak erosion rate at 60° impact angle.

**Table 3.3    Abrasion resistant elevated temperature strength alloys**

| Alloy | Characteristic |
|---|---|
| 1.  Haynes 6B (wrought) | |
| 2.  Stellite No. 6 (oxyacetylene deposit) (Co-30Cr-4W-1.1C) | 6B and No. 6 have similar compositions, but very different microstructures. 6B has large discrete carbides and an elongation of 11%; No. 6 has a cast microstructure and 1% elongation |
| 3.  Tristelle alloys TS-1, 2, 3. (Fe-35Cr-12Co-20Ni-1, 2, or 3C) | Series of iron-base wear-resistant alloys containing up to 3% C: Oxyacetylene-deposited TS-1 may be compared to Stellite No. 6: similar carbide volume, microstructure, and elongation |
| 4.  Ferralium 255 (30% cold reduced) (Fe-26Cr-5Ni-3Mo) | High-strength duplex stainless steel that has high strain-hardening coefficient |
| 5.  Haynes 188 (annealed and 30% cold reduced) (Co-22Ni-22Cr-14W) | Cold reduction reduces elongation from 50 to 10% |
| 6.  Hastelloy C-276 (annealed and 30% cold reduced) (Ni-16Cr-16Mo-5Fe-4W) | Cold reduction reduces elongation from 65 to 15% |
| 7.  Cabot 718 (annealed and aged) (Ni-19Cr-19Fe-3Mo-5(Nb+Ta)) | Aging doubles yield strength and reduces elongation from 30 to 12% |
| 8.  Berylco 25 (annealed and aged) (Cu-1.85Be) | Aging doubles yield strength and reduces elongation from 38 to 4% |

**Table 3.4    Erosion rates of abrasion resistant and heat resistant alloys**

| Alloy | Elongation, % | Hardness, HV | Erosion rate, $cm^3/g \times 10^{-5}$ Silicon carbide erodent 30° | 60° | 90° | Quartz erodent 30° | 60° | 90° |
|---|---|---|---|---|---|---|---|---|
| Haynes 6B (wrought) | 11 | 494 | 2.29 | 2.26 | 1.94 | 1.14 | 1.28 | 1.29 |
| Stellite No. 6 (deposit) | 1 | 465 | 2.32 | 2.36 | 1.96 | 1.08 | 1.48 | 1.34 |
| Tristelle TS-1 (deposit) | 1 | 330 | 2.74 | 2.55 | 2.10 | 1.50 | 1.60 | 1.47 |
| Tristelle TS-2 (wrought) | ... | 430 | ... | ... | ... | 1.46 | 2.11 | 1.87 |
| Tristelle TS-3 (wrought) | ... | 460 | ... | ... | ... | 1.60 | 2.45 | 1.91 |
| Ferralium 255 (cold reduced) | 22 | 407 | 2.31 | 1.81 | 1.45 | 1.38 | 1.31 | 1.08 |
| Haynes 188 annealed | 53 | 320 | 2.07 | 1.89 | 1.54 | 1.07 | 1.41 | 1.16 |
| Haynes 188 cold reduced | 9 | 465 | 2.18 | 1.89 | 1.52 | 1.18 | 1.38 | 1.16 |
| Hastelloy C-276 annealed | 67 | 229 | 2.10 | 1.78 | 1.46 | 1.20 | 1.22 | 1.07 |
| Hastelloy C-276 cold reduced | 15 | 408 | 2.13 | 1.77 | 1.52 | 1.17 | 1.32 | 1.11 |
| Cabot 718 annealed | 30 | 243 | 2.15 | 1.75 | 1.54 | 1.22 | 1.29 | 1.10 |
| Cabot 718 aged | 12 | 476 | 2.28 | 1.93 | 1.59 | 1.32 | 1.40 | 1.17 |
| Berylco 25 annealed | 38 | 199 | 2.81 | ... | ... | ... | ... | ... |
| Berylco 25 aged | 4 | 410 | 3.07 | ... | ... | ... | ... | ... |

Room temperature erosion; air carrier gas; 60 m/s impingement; 250 μm silicon carbide; 135 μm quartz

The volume-loss data in Table 3.4 shows that the erosion rates for these alloys, which cover a wide range of mechanical properties, are all quite similar, and that the variation in erosion rate with angle is small. There are no great differences between iron-base, cobalt-base, or nickel-base alloys. With the exception of the higher carbon content Tristelle alloys, the maximum erosion rates were no more than approximately 40% greater at the peak impact angle than the rates at 90° impact angle. Classically ductile materials typically exhibit a factor of 3 in their erosion rates between maximum and 90° impact angles, while classically brittle materials display a continuously increasing erosion rate with angle (Ref 15).

The relationships between indentation hardness and erosion rate, and between ductility and erosion rate are shown in Fig. 3.12 and 3.13, respectively. Where the hardness or ductility has been changed by thermal or mechanical treatment, points have been plotted to show the erosion rate for the material in both the annealed and treated conditions, and those points joined by a solid line. The lines are for clarity of presentation only; they are not meant to imply an attempt to interpolate between pairs of data points. However, the slopes of the lines do indicate the trends in erosion rate that changes in elongation or hardness have on particular alloys. Because the aging or cold-working treatments simultaneously change a number of mechanical properties, the hardness and elongation are not independently altered. It can be seen that only small changes in erosion rate occur for large changes in elongation and hardness.

Elongation and hardness have a roughly inverse relationship during thermomechanical treatments; increasing one almost invariably re-

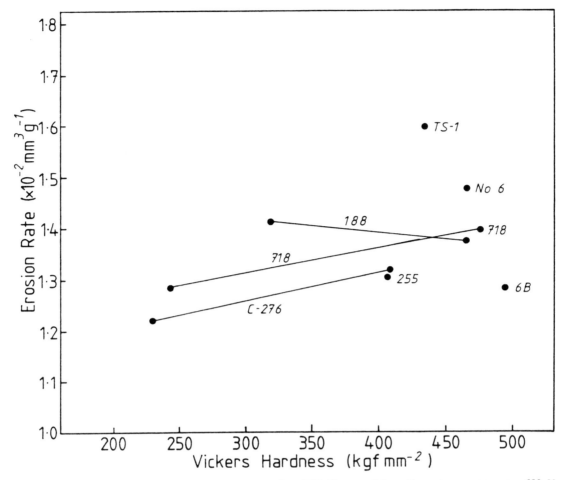

**Fig. 3.12** Erosion rate versus Vickers hardness number (HV). Test conditions: Room temperature, $\alpha = 60°$, $V$ = 60 m/s, 135 μm quartz particle

duces the other. To have erosion performance of these types of alloys changed so little by large changes in hardness and ductility is somewhat different from the behavior of low-alloy steels discussed earlier. Figures 3.12 and 3.13 indicate that neither elongation nor indentation hardness dominated erosion behavior at the expense of the other. Thus, different types of structural alloys with relatively large property differences can have similar erosion behavior.

Examination of eroded surfaces by scanning electron microscopy (SEM) confirms the indications of the mass loss data. For a given erodent, the appearance of the eroded surfaces are very similar for a wide variety of alloys that have large differences in their mechanical properties. No

differences were observed between the eroded surfaces of annealed alloys and those that had been aged or cold worked. It was hoped that quartz, being a milder erodent than silicon carbide, would differentiate between the alloys. While this was true for the mass loss measurements, the morphology differences of the eroded surfaces were too small to be detected. A typical example is shown in Fig. 3.14, which compares cold-worked and annealed Hastelloy C-276.

Figure 3.15 compares the surfaces of Stellite No. 6, Haynes 6B and Tristelle TS-1 eroded by silicon carbide at 30° impact angle. Stellite No. 6 (cobalt-base), and Tristelle TS-1 (iron-base) were both oxyacetylene deposited, resulting in the formation of interdendritic carbides during solidifi-

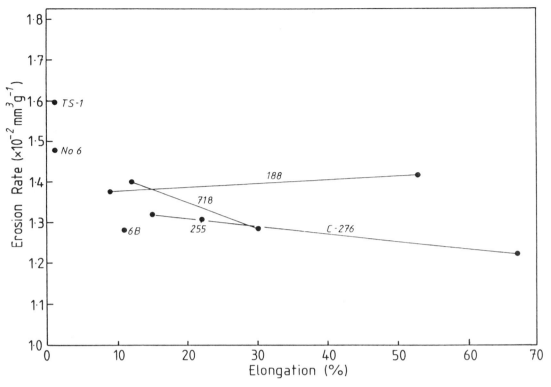

**Fig. 3.13** Erosion rate versus percentage elongation in a tensile test. Test conditions: room temperature, $\alpha =$ 60°, $V = 60$ m/s, 135 μm quartz particle

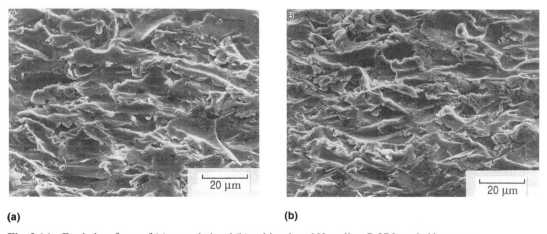

**(a)**  **(b)**

**Fig. 3.14** Eroded surfaces of (a) annealed and (b) cold-reduced Hastelloy C-276 eroded by quartz

cation. Haynes 6B is a wrought alloy having a composition close to that of Stellite No. 6 but with much more rounded carbides dispersed uniformly throughout the matrix; its tensile elongation is ten times that of the deposited alloys. Although the eroded surfaces show many similar features, it is evident that Tristelle TS-1 has more deformed material attached to its surface, while the cobalt-base alloys have a "cleaner" surface. The difference in erosion rates between Haynes 6B and Stellite No. 6 is very small, much smaller than the difference between Tristelle TS-1 and

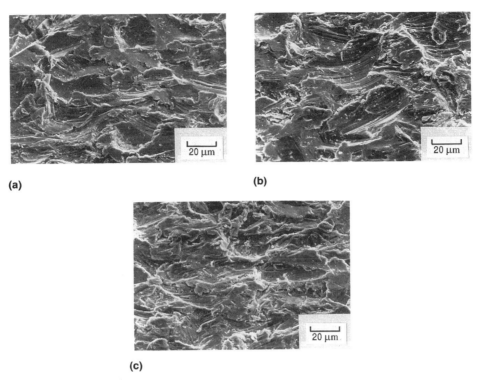

**Fig. 3.15** Eroded surfaces of (a) Stellite 6, (b) Haynes 6B, and (c) Tristelle TS-1 eroded by SiC particles

Stellite No. 6 (Table 3.4). These factors indicate that the morphology of the carbides has a much smaller effect on erosion performance than does the composition of the matrix containing them. The intrinsic hardness and ductility of the matrix is more important than the hardness and ductility of the whole alloy.

## Correlation of Erosion Conditions, Erosion Rates, and Surface Morphologies(Ref16)

The erosion of metals by small solid particles is achieved by a high strain rate extrusion/forging/fracture mechanism that differs from other types of wear processes by its near independence from most of the mechanical properties of metals associated with other types of wear behavior, that is, abrasive wear and sliding wear behavior. Erosive wear does depend strongly on the conditions under which it occurs. Variables such as the velocity and impact angle of the erodent particles,

as well as particle properties such as shape, size, and fracture strength, have major effects on material wastage. It is important to understand how test conditions can have a major effect on erosion rates while the target materials' properties do not. A series of tests was performed to determine how and why test variables have such a strong effect on erosion. Correlation of the test variables, the resulting erosion rates, and the eroded surface morphologies provides the necessary understanding.

### Erosion Rates

The erosion rates of the alloys eroded with SiC particles are listed in Table 3.5. The steady-state erosion of the alloys eroded with smaller-size $Al_2O_3$ particles are listed in Table 3.6. All weight losses were normalized to account for the different densities of the alloys tested by dividing the weight loss by the density of the alloy. This resulted in the determination of the volume of material loss per gram of eroding particles.

It can be seen in Table 3.5 that the erosion rates increased with particle velocity and were higher at the lower impact angle. More volume of material was eroded from the aluminum than from the steel alloys. The two steel alloys eroded at a similar rate except at the highest velocity where the mild steel eroded at a lower rate than the 304 stainless steel. This difference at the highest velocity used in the experiment revised a trend observed at lower velocities that was discussed earlier in this chapter where 304 stainless steel eroded at a lower rate than 1020 steel calculated on a gram-per-gram (g/g) basis (Ref 2).

In Table 3.6, the erosion rates for the three steels tested using the $Al_2O_3$ erodent were more similar at all of the particle velocities. At the 60 and 90 m/s velocities, the erosion rates were lower than for the same alloys eroded with SiC particles. This behavior reflects the effect of the erodent particle shape on metal wastage; the somewhat more angular shape of the SiC particles compared to the $Al_2O_3$ particles. The effects of particle variables on erosion will be discussed in detail in Chapter 4. However, an incongruity exists at the lowest test velocity. At 30 m/s, the $Al_2O_3$ particles removed more volume of mate-

**Table 3.5  Erosion rates of steel and aluminum alloys using silicon carbide erodent**

| Material | Hardness, HV | Erosion rate, $cm^3/g \times 10^{-5}$ | | | | | | | |
|---|---|---|---|---|---|---|---|---|---|
| | | $\alpha = 30°$ | | | | $\alpha = 90°$ | | | |
| | | 30 m/s | 60 m/s | 90 m/s | 130 m/s | 30 m/s | 60 m/s | 90 m/s | 130 m/s |
| 1020 steel | 135 | 0.86 | 4.10 | 9.10 | 9.50 | 0.32 | 1.75 | 4.54 | 5.50 |
| 304 stainless steel | 198 | 0.72 | 3.53 | 9.97 | 13.10 | 0.38 | 2.31 | 7.50 | 10.40 |
| 1100 aluminum | 42 | 1.6 | 6.3 | 13.3 | 19.34 | 0.94 | 3.40 | 9.92 | 12.40 |

300 g of 250 μm silicon carbide particles

**Table 3.6  Erosion rates of steel alloys using $Al_2O_3$ erodent**

| Material | Hardness, HV | Erosion rate, $cm^3/g \times 10^{-5}$ | | | | | |
|---|---|---|---|---|---|---|---|
| | | $\alpha = 30°$ | | | $\alpha = 90°$ | | |
| | | 30 m/s | 60 m/s | 90 m/s | 30 m/s | 60 m/s | 90 m/s |
| 1020 steel | 135 | 0.46 | 1.31 | 2.82 | 0.28 | 0.76 | 1.33 |
| 304 stainless steel | 198 | 0.25 | 1.15 | 3.11 | 0.12 | 0.76 | 2.22 |
| 4340 | 290 | 0.50 | 1.80 | 3.02 | 0.25 | 1.23 | 2.51 |

300 g of 100 μm $Al_2O_3$ particles

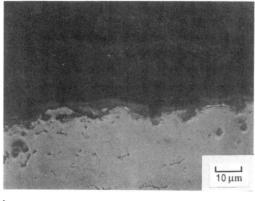
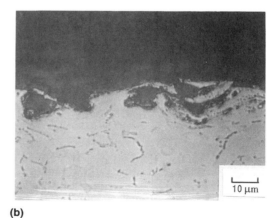

(a)                                                    (b)

**Fig. 3.16**   Cross section of 1020 steel eroded by 250 μm SiC particles. Particle impact angle: $\alpha = 90°$. Test velocity: (a) $V = 30$ m/s, and (b) $V = 130$ m/s

rial than the SiC particles for the 1020 and 304 stainless steels.

### Microstructures of Eroded Surfaces

All of the specimens tested at all conditions using SiC erodent had the typical platelet formations on their surfaces described in Chapter 2 (Ref 1). The 1020 and 304 stainless steels had essentially the same type of surface appearance at all test conditions. The texture of the eroded surface consisted of shallow craters and platelets at various stages of development and sizes. The effect of the erodent particle velocity is seen in Fig. 3.16. The sizes of the craters and platelets on the 1020 steel specimen eroded at 130 m/s are considerably larger than those eroded at 30 m/s. The increased kinetic energy of the particles moving at the higher velocity caused deeper craters and larger platelets to form. Table 3.5 shows the effect of the larger platelets on the erosion rate. An increase in the 1020 steel erosion rate of 11 times occurred between the 30 m/s and 130 m/s particle velocities at 30° impact angle and 17 times at 90° impact angle.

Figures 3.17 and 3.18 show the surface and cross section of the 304 stainless steel eroded at velocities of 30 and 130 m/s. The principal difference in surface appearance between the low and high velocities is the size of the craters in the 30° impact angle tests, being somewhat longer in the 130 m/s specimens. The basic orientation of the craters of the 30° impact angle eroded specimens is along the direction of the impacting particles as shown by the arrows on the figure. There is no strong orientation of the platelets. There is no readily apparent difference in the surface appearance of the 90° impact specimens at the two test velocities.

Comparing the 30 and 90° impact specimens shows that the same platelet mechanism occurs at both angles. Larger craters are seen on the 90° impact specimen compared to the 30° impact specimen at the low velocity. At the 130 m/s velocity, there is no distinct difference, either in the surface photos in Fig. 3.17 or in the cross-section photos in Fig. 3.18. This relates to the relatively small difference in the erosion rates of the two specimens listed in Table 3.5.

Figure 3.19 shows the cross section of the eroded surface of the 1100-O aluminum at the same magnification for the specimens eroded at $V = 30$ and 90 m/s. As was observed for the 1020 steel, there is a major difference in the size of the platelets and the depth of the craters. This difference accounts for the order of magnitude increase in the erosion rate from $V = 30$ m/s to $V = 90$ m/s.

Comparing the eroded surface cross sections of the 1020 steel and 1100-O aluminum (Fig. 3.16, 3.19), shows the effect of higher velocity particles on the softer, more extrudable 1100-O aluminum. The sizes of the craters and platelets from the higher velocity tests are much larger in the aluminum specimen than in the 1020 steel specimen. Comparing the two materials at the lower velocity, 30 m/s, there does not appear to be much difference in the texture of the two surfaces. It is to be noted that a classic-shaped platelet is shown in Fig. 3.16(a) on the left side. Its cross section is similar to that of a flat mushroom with a stem that attaches the platelet to the base metal.

## Incremental Erosion Rates[Ref 16]

The nature of the incremental erosion rate curve near the time of initiation of erosion was investigated to determine what occurs when the early platelets are being formed on the eroding surface. The velocity of the particles as well as their impact angle and size were varied. In previous work (Ref 1, 9), incremental erosion rate curve determinations were made using 10 to 60 g increments of erodent. In this study, increments of 1 to 2 g of erodent were used for the first 10 g of erodent and 5 g increments were used up to 30 g.

### Effect of Impact Angle

It was determined that an incremental erosion rate peak occurred at all test conditions near 20 g of particles. Figure 3.20 plots the incremental erosion rates of 1018 steel at three impact angles: 20, 30, and 90°. The peak rate can be seen on all three curves at about 20 g of particles. The peak rate is followed by a sharp decrease in the rate to

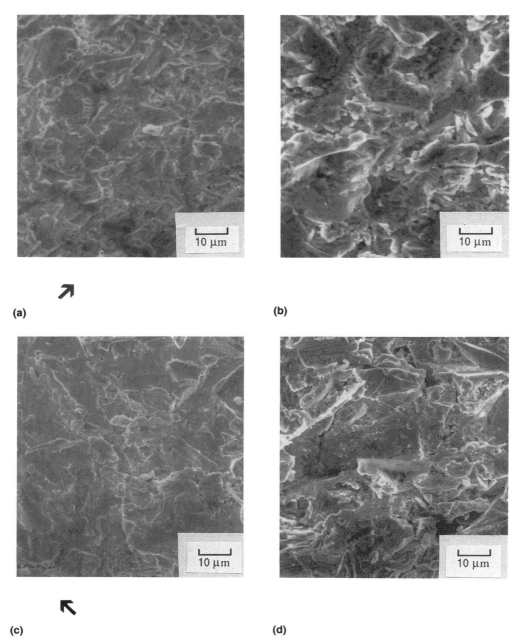

**Fig. 3.17** Surface of 304 stainless steel eroded by 250 μm SiC particles. (a) $\alpha = 30°$, $V = 30$ m/s, Erosion rate $= E = 0.72 \ 10^{-5}$ cm$^3$/g. (b) $\alpha = 90°$, $V = 30$ m/s, $E = 0..38 \ 10^{-5}$ cm$^3$/g. (c) $\alpha = 30°$, $V = 130$ m/s, $E = 13.10 \ 10^{-5}$ cm$^3$/g. (d) $\alpha = 90°$, $V = 130$ m/s, $E = 10.40 \ 10^{-5}$ cm$^3$/g.

a level slightly below the steady-state level and then an increase in rate up to the steady-state rate. The steady-state erosion rates are highest at 30° and lowest at 90° impact angles, as expected. The consistency of the incremental erosion rates in the steady-state region is evident in all three curves. The general shape of all three curves is representative of the platelet mechanism of erosion wherein an initial threshold period occurs with no measurable erosion, followed by an increasing

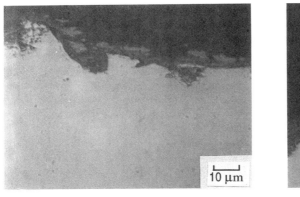 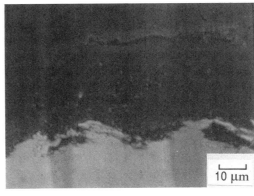

**(a)**        **(b)**

**Fig. 3.18**   Cross section of 304 stainless steel eroded by 250 μm SiC particles. (a) α = 30°, V = 130 m/s. (b) α = 90°, V = 130 m/s

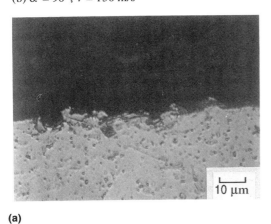 

**(a)**        **(b)**

**Fig. 3.19**   Cross section of 1100 aluminum eroded by 250 μm SiC particles. (a) α = 30°, V = 30 m/s. (b) α = 30°, V = 90 m/s

rate up to the initial peak rate and then down to a final steady-state rate (Ref 17). It is interesting to note that even at shallow impact angles that are less than the peak erosion rate angle, the shape of the curve is the same. This indicates that the platelet mechanism is still the active mechanism. It was thought that possibly at the shallow impact angles, the microcutting mechanism might have become the major mechanism.

Figure 3.21 further verifies that platelet formation dominates at less than the peak impact angle. It also contributes to understanding why there is a peak erosion rate near 20 g of particles. Both photos show the presence of platelets and shallow craters in tests carried out at a 20° impact angle. However, there is an observable difference in the

surface texture of Fig 3.21(a), which was taken at the peak erosion rate shown in Fig. 3.20, and Fig. 3.21(b), which was taken at steady-state erosion. The surface at the peak incremental erosion rate consists of relatively large craters and platelets compared to the smaller size of the craters and platelets at the steady-state erosion rate. A stereo pair of each of the surfaces showed the larger platelets at peak erosion to be protruding much higher above the surface than the smaller platelets at steady-state erosion. Thus, at the peak erosion condition, prior to reaching steady-state erosion, the platelets are both larger and more vulnerable to being knocked off the surface than are the platelets at steady-state erosion. This same pattern was also observed on specimens tested at 30 and 90° impact angles.

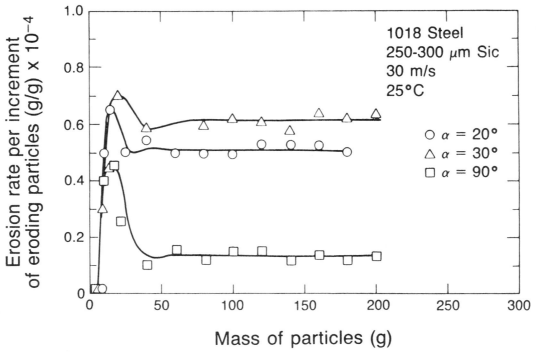

**Fig. 3.20** Incremental erosion rates of 1018 steel at α = 20, 30, and 90°

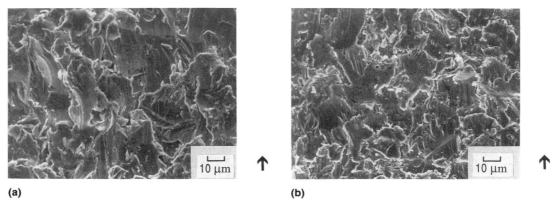

**(a)**                                                    **(b)**

**Fig. 3.21** Surface of eroded 1018 steel at (a) incremental peak and (b) steady-state erosion rates. α = 20°, V = 30 m/s. 250 to 300 μm SiC. T = 25 °C

As erosion approaches the steady-state rate and the full texture of the eroding surface is achieved, some platelets are mashed down into nearby craters, thereby decreasing their vulnerability and making them more difficult to remove from the surface. The number that are knocked off the surface by an increment of eroding particles thereby decreases. The mashed-down platelets will generally be removed as the result of their fracture into pieces by the impacting particles rather than the whole platelet being removed by the fracture of its attachment stem as occurs more often near the beginning of the erosion process. An example of the latter configuration is shown in Fig. 3.16(a), which shows a well-formed platelet with its attachment stem intact. This sequence of events explains the shape of the curves in Fig. 3.20.

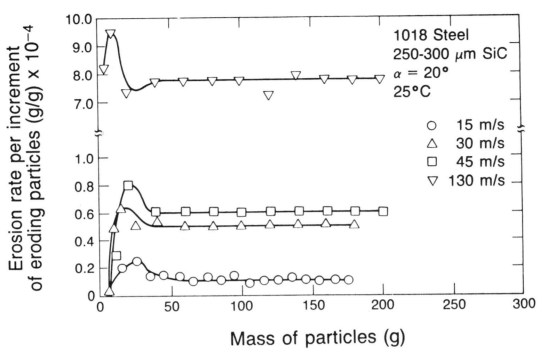

**Fig. 3.22**  Incremental erosion rates of 1018 steel at $V$ = 15, 30, 45, and 130 m/s

**Table 3.7  Carbide-metal materials**

| Alloy | Nominal composition, wt% | Carbide, vol% |
|---|---|---|
| Pure iron | Ingot iron | 0 |
| Spheroidized 1020 | Fe-0.2C | 3.1 |
| Spheroidized 1080 | Fe-0.8C | 12.0 |
| Stellite No. 6 | Co-30Cr-4W-1.1C | 10.4 |
| Haynes 6B | Co-30Cr-4W-1.1C | 10.4 |
| Tristelle TS-1 | Fe-30Cr-10Ni-12Co-5Si-1C | 8.3 |
| Tristelle TS-2 | Fe-35Cr-10Ni-12Co-5Si-2C | 16.8 |
| Tristelle TS-3 | Fe-35Cr-10Ni-12Co-5Si-3C | 21.4 |
| Cr-Mo white iron | Fe-20Cr-2.5Mo-2.6C | 23.1 |
| K90 | WC-25Co | 63.1 |
| K3520 | WC-20Co | 69.5 |
| K3055 | WC-10Co | 83.7 |
| K701 | WC-10Co-4Cr | 76.2 |
| K801 | WC-5.7Ni-0.4Co | 89.9 |
| K162B | TiC-6NbC-2WC-25Ni-7Mo | 78.5 |
| K165 | TiC-7WC-2.5NbC-9Ni-9Mo | 88.8 |

## Effect of Velocity on Erosion Mechanism

A series of tests was carried out at a 20°, shallow impact angle at several velocities to determine whether the predominant erosion mechanism of platelet formation and removal would be replaced by a microcutting mechanism as the velocity increased. Figure 3.22 shows that the incremental erosion rate curves for 1018 steel had the same pattern at particle velocities of 15, 30, 45, and 130 m/s. The initial peak erosion rates occurred near 20 g of particles, the same as is shown in Fig. 3.20. The steady-state erosion rates were constant and related directly to the particle velocity. Even at the high velocity of 130 m/s, the pattern of the incremental erosion rate curve was the same.

## Erosion of Carbide-Containing Metals[Ref 18]

Carbide particle-metal binder composites are commonly used materials to resist severe wear conditions in such applications as cutting tools, rotating seals, slurry valves, and other kinds of components that are subjected to sliding or abrasive wear at high contact forces. Correlations between composite hardness and wear resistance are possible for sliding and abrasive types of wear. However, solid-particle erosive wear does not conform to this relationship. In order to ascertain whether carbide composites have any basis for selection in erosive-wear applications, an investigation was carried out that used materials with carbide contents ranging from 0 to 89 vol% carbide in metal matrices. Several methods were used to generate and distribute the carbide constituents in the generally ductile metal matrices, resulting in a number of different morphologies and carbide compositions.

### Materials and Test Conditions

The alloys tested are listed in Table 3.7, together with their nominal compositions and the volume percentage of carbide they contain. The AISI 1020 and 1080 steels were spheroidized by quenching into water from an austenitizing treatment, followed by holding at 705 °C for 72 h. This produced a uniform dispersion of carbides 1.5 to 2 µm in diameter in a ferrite matrix. The pure iron was hot rolled to a thickness suitable for

testing and was, therefore, somewhat work hardened. The series of Tristelle alloys and the Cr-Mo white cast iron all had as-cast microstructures. Tristelle TS-1 and the white cast iron were hypoeutectic, Tristelles TS-2 and TS-3 were hypereutectic. Only TS-3 contained an appreciable quantity of large primary carbides that were 100 to 200 µm in length, whereas the eutectic carbides were of the order of 10 µm in length. Stellite 6 (oxyacetylene-deposited) had a cast-type microstructure with carbides 10 to 15 µm in length while Haynes 6B was a wrought alloy of similar composition but with much more rounded carbides. The K series cermets were used as supplied and had carbide grain sizes in the range 1 to 10 µm.

Erosion tests were carried out at room temperature using a nozzle tester with 75 to 200 µm angular quartz erodent at 60 m/s velocity. Tests were conducted at impact angles of 30, 60, and 90°. Quartz sand was used as the erodent in order to match reasonably closely the hardness of abrasives found in service situations. This is felt to be important when testing carbide-containing, composite materials.

### Material Loss

The steady-state erosion volume losses are presented in Table 3.8 and the losses incurred at a 60° impact angle are plotted against volume percentage of carbide in Fig. 3.23. The plain-carbon steels and pure iron can be considered to form a comparable group of materials, as do the

**Table 3.8  Erosion of carbide-metal materials**

| Alloy | Carbide, vol% | Volume erosion rate (m³/g × 10⁻¹²) | | |
| --- | --- | --- | --- | --- |
| | | 30° | 60° | 90° |
| Pure iron | 0 | 9.70 | 7.76 | 5.77 |
| 1020 | 3.1 | 10.05 | 7.86 | 5.67 |
| 1080 | 12.0 | 9.99 | 8.19 | 6.22 |
| Stellite No. 6 | 10.4 | 10.8 | 14.8 | 13.4 |
| Haynes 6B | 10.4 | 11.4 | 12.8 | 12.9 |
| Tristelle TS-1 | 8.3 | 11.78 | 12.18 | 11.51 |
| Tristelle TS-2 | 16.8 | 14.43 | 20.87 | 18.54 |
| Tristelle TS-3 | 21.4 | 15.94 | 24.37 | 19.02 |
| White iron | 23.1 | 10.61 | 13.33 | 12.27 |
| K90 | 63.1 | 7.04 | 11.47 | 12.12 |
| K3520 | 69.5 | 4.83 | 10.21 | 10.73 |
| K3055 | 83.7 | 2.21 | 4.70 | 4.10 |
| K701 | 76.2 | 0.11 | 0.10 | 0.20 |
| K801 | 89.9 | 0.46 | 0.77 | 0.61 |
| K162B | 78.5 | 5.33 | 10.29 | 12.59 |
| K165 | 88.8 | 1.23 | 2.51 | 1.95 |

Stellite-Tristelle alloys. The K series cermets form a third group. Viewed in this way it is evident that a broad trend exists whereby increasing the carbide volume fraction increases erosion rates at low volume fractions but decreases them at the high fractions where the carbide is the major constituent. Because of the differences that exist between the carbide shapes and sizes and the matrix compositions of the three groups, emphasis is placed upon comparisons within each group rather than between the groups. Within each group, the microstructures are sufficiently similar that the major variable is carbide volume fraction. Plots made using the 30 or 90° impact angle

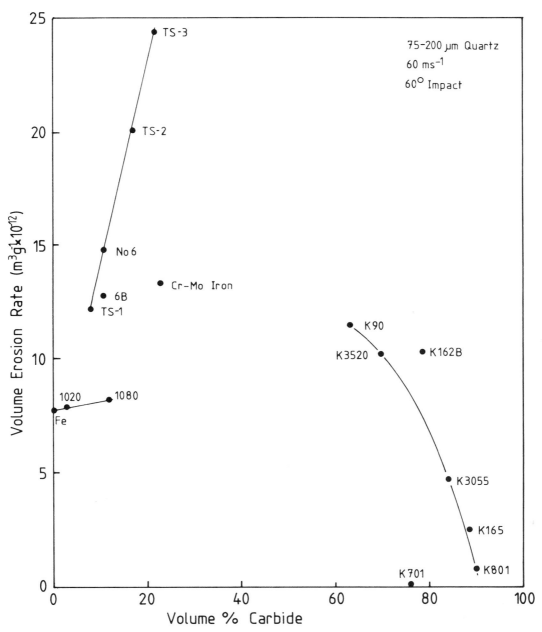

**Fig. 3.23** Erosion rates of materials as a function of carbide content

results show similar trends, but the relative positions of points are changed because the lowest carbide content alloys behave in a classically ductile manner while other alloys behave in a more brittle manner. It is convenient to discuss separately the behavior of low carbide content alloys and high carbide content materials (cermets).

**Low Carbide Content Alloys.** The low carbide fraction alloys consist of carbides dispersed throughout a ductile matrix with some variation between the alloys in the size, shape, and distribution of carbides. It is generally the case that the carbides cause plastic flow of the matrix to be inhomogeneous owing to their high hardness and stiffness. This is clearly shown in Fig. 3.24, where the very high strains that must take place in the matrix close to a carbide result in the formation of localized voids in the matrix. The associated high local stresses also cause nearby carbide fracture. It should be noted that Stellite 6 has an inhomogeneous microstructure; the carbides are concentrated in the eutectic mixture between the proeutectoid dendrites of the cobalt-base matrix. The area shown in Fig. 3.24 is such a eutectoid region. The extensive damage that occurs in regions of high strain, such as at the crater lip results in deformed regions being particularly vulnerable to removal during subsequent impacts.

Considering the alloy group pure iron-1020-1080 steels, it is evident from Fig. 3.23 that the presence of 12 vol% carbide (the level in 1080 steel) has a mildly deleterious effect on the erosion rate compared with that of pure iron. The erosion rate has increased by 5.5% at 60° impact and by 8% at 90° impact. Examination of specimens eroded by quartz does not reveal extensive void formation or carbide fracture. The eroded surface is covered with platelets and craters. Eroded pure iron and 1020 steel specimens had similar features to eroded 1080 steel. In particular, there were numerous occurrences of embedded abrasive and platelets folded over into adjacent craters.

The rather small size of the carbides in spheroidized 1020 and 1080 steels results in lower strain gradients than is the case when the same carbide volume fraction is present but the carbides are larger. Less disruption of plastic flow occurs and thus less damage accumulates at the carbides (Ref 9). While the comparison is not ideal given the different matrix compositions, a comparison between Haynes 6B and 1080 steel, which have about the same carbide content, illustrates the effect. Haynes 6B has a slightly smaller carbide volume fraction than 1080 steel (in both cases the distribution is homogeneous), but the larger carbide size in Haynes 6B results in more damage at the carbides owing to higher strain gradients with a resulting greater erosion loss.

Erosion rates increase sharply for the Tristelle alloys as the carbide content increases, see Fig. 3.23. It is considered coincidental that the data points for Stellite 6 and Haynes 6B lie so close to the line through the Tristelles; only alloys of very similar composition and processing route can be expected to show such systematic behavior changes. These alloys, including white cast iron, all have relatively large carbides. With the exception of Haynes 6B, the carbide distribution is also inhomogeneous.

Deformation under an impacting particle in the Tristelles is forced to be localized by the carbide network. Such localization, and the low ductility in a region of high carbide content, allows detachment of displaced material after a smaller number of impacts than would be the case in a single-phase alloy. In the case of the higher carbide fraction alloys, TS-3 and the Cr-Mo white iron, a number of carbides are quite large. Under these circumstances, the ductile matrix lying above a carbide is particularly vulnerable. The carbide acts as an anvil and deformation of the matrix is particularly localized and severe. Once exposed, the carbide is unsupported and is itself vulnerable to fracture either through its whole section or at its edges (Ref 19).

**K Series Cermets.** While the low carbide fraction alloys consist of carbides dispersed (not necessarily uniformly) throughout a ductile matrix, the cermets are better described as aggregates of carbides in which the gaps between carbides are filled with binder material. For the high carbide fraction cermets, the carbide network is largely continuous owing to dissolution and subsequent precipitation of carbide during sintering and cooling. The carbide network can then be considered to form a skeleton. The binder provides a mechanical, three-dimensional clamp on the particles as well as being chemically bound to them. Because the carbide content is so high for nearly all the cermets tested, plasticity on the scale of a typical erosion crater is not possible. There is insufficient binder to allow the carbides to move past each other.

The only exception to this is K90 (63 vol% carbide). Scanning electron micrographs of eroded K90 and K3055 (84 vol% carbide), both WC-Co cermets, are shown in Fig. 3.25. It is evident that material is lost from K90 (Fig. 3.25a)

by a mechanism similar to that which occurs when much softer and more ductile alloys are eroded, that is, material is extruded and subsequently forged before being detached from the surface. The volume removed in such a sequence of events may contain many carbides; these can aid material removal on account of the reduction in strain to failure that they cause. Because large pieces may be lost at a given time, the erosion rate of K90 is high compared with the other cermets (Table 3.8).

In contrast to K90, the eroded surface of K3055 (Fig. 3.25b) exhibits neither impact craters nor any other features to indicate that signifi-

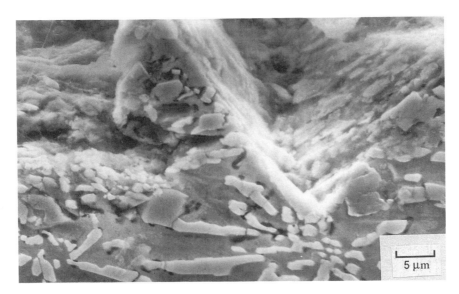

**Fig. 3.24**   Taper section of Stellite 6 eroded area

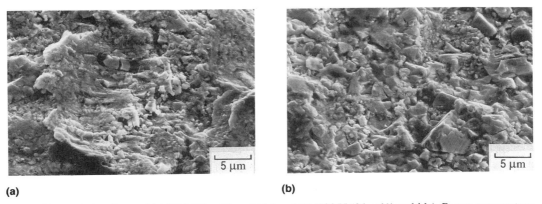

**(a)**                                                                 **(b)**

**Fig. 3.25**   Eroded surfaces of (a) K90 (63 vol% carbide) and (b) K3055 (84 vol% carbide). Room-temperature tester; air-carrier gas; 135 μm (average) quartz particles. $\alpha = 60°$, $V = 60$ m/s, $T = 25\ °C$

**Fig. 3.26** Taper sections of (a) K162B and (b) K165 cermets. Room-temperature tester; air-carrier gas; 135 μm (average) quartz particles. $\alpha = 60°$, $V = 60$ m/s, $T = 25$ °C

cant plasticity has taken place because, with only 17 vol% binder and a high carbide contiguity, the carbide skeleton is rigid. The eroded surface is almost entirely covered with exposed carbide grains; a number of these, particularly the larger grains, have been fractured. Because the erodent used was of the order of 30 times larger than the carbides in K3055, it is unlikely that any binder was preferentially removed from between carbides to any significant depth.

The probable erosion mechanism for cermets of high carbide volume fraction, approximately 80 vol% and greater, is one in which individual carbide grains or pieces of them are removed by impacts. Cracks introduced into the carbide skeleton might also allow small clusters of grains to be removed together. Fractures in larger carbide grains have been observed that could result in the loss of small chips of carbide by subsequent impacts. For cermets of less than 80 vol% carbide, the binder controls erosion. Gross deformation involving crater formation need not take place for the binder to be the controlling phase, although this does occur for lower carbide fraction cermets (Fig. 3.24). For cermets where the carbide skeleton dominates the erosion behavior, a carbide volume fraction of at least 80% is required based upon consideration of measured erosion rates as the carbide volume fraction was varied (Ref 20).

The trend of decreasing erosion with decreasing binder content is followed smoothly by the TiC-Ni-Mo binder cermet K165 and the WC-Ni

binder cermet K801 as well as the family of WC-Co binder cermets K90, K3520, and K3055. Because the erosion of K165 and K801 are both controlled by the carbide, it is possible that such carbide control is dominated by the geometry of the skeleton and the carbide contiguity rather than by particular cermet compositions/mechanical properties. For example, nickel forms a weaker bond with tungsten carbide than does cobalt (Ref 20), and titanium carbide will have different mechanical properties from tungsten carbide. Thus, if the binder-carbide bond strength or the carbide fracture toughness were important parameters in determining carbide-dominated erosion, it would not be expected that K165 and K801 would lie on the line through data points for WC-Co alloys (K90, K3520, and K3055).

Results for two of the cermets tested depart significantly from the line through the other points in Fig. 3.23. K162B erodes more than might be expected and K701 less than might be expected from the other data. Like K165, K162B is a Ni-Mo binder titanium-carbide alloy but has sufficient binder to have it control erosion. Scanning electron micrographs of taper sections through K165 and K162B are shown in Fig. 3.26. The higher binder content of K162B manifests itself in the greater erosion deformation appearance of the eroded surface. K701 is a Co-Cr binder tungsten carbide alloy (76 vol% carbide) of quite fine grain size that results in smaller size carbide pieces being eroded off.

## References

1. Levy, A.V., The Erosion of Metal Alloys and Their Scales, *Proc. NACE Conf. Corrosion-Erosion-Wear of Materials in Emerging Fossil Energy Systems* (Berkeley, CA), 1982, p 298-376

2. Foley, T., and Levy, A.V., The Erosion of Heat Treated Steels, *Wear*, Vol 91 (No. 1), 1983, p 45-64

3. Levy, A.V., and Chik, P., The Effects of Erodent Composition and Shape on the Erosion of Steel, *Wear*, Vol 89 (No. 2), 1983, p 151-162

4. Gulden, M.E., and Kubarych, K.G., "Erosion Mechanisms of Metals," Report SR 81-R4526-02, Solar Turbines, Inc., San Diego, CA, 1982

5. Gulden, M.E., Influence of Brittle to Ductile Transition on Solid Particle Erosion Behavior, *Proc. 5th Int. Conf. Erosion Liquid and Solid Impact* (Cambridge, England), Sept 1979

6. Christman, T., and Shewmon, P., Erosion of a Strong Aluminum Alloy, *Wear*, Vol 52 (No. 1), Jan 1979, p 57-70

7. Shewmon, P., Particle Size Threshold in the Erosion of Metals, *Wear*, Vol 68 (No. 2), May 1981, p 253-258

8. Naim, M., and Bahadur, S., Effect of Microstructure and Mechanical Properties on the Erosion of 18 Ni (250) Maraging Steel, *Proc. ASME Int. Conf. Wear of Materials* (Vancouver, B.C., Canada), 1985, p 586-594

9. Levy, A.V., The Solid Particle Erosion of Steel as a Function of Microstructure, *Wear*, Vol 68 (No. 3), 1981, p 269-287

10. Suh, N., Special Issue: Delamination Wear and Ferrography, *Wear*, Vol 44 (No. 1), Aug 1977, p 1-162

11. Jahanmir, S., The Mechanics of Subsurface Damage in Solid Particle Erosion, *Wear*, Vol 61 (No. 2), June 1980, p 309-324

12. Brown, R.; Jun, E.; and Edington, J., Mechanisms of Solid Particle Erosion Wear for 90° Impact on Copper and Iron, *Wear*, Vol 74 (No. 1), Dec 1982, p 143-156

13. Brown, R., and Edington, J., Erosion of Copper Single Crystals Under Conditions of 90° Impact, *Wear*, Vol 69 (No. 3), July 1981, p 369-382

14. Ninham, A., The Effect of Mechanical Properties on Erosion, *Wear*, Vol 121 (No. 3), 1988, p 307-324

15. Rickerby, D.G., and MacMillan, N.H., Erosion of Aluminum and Magnesium Oxide by Spherical Particles, *Proc. ASME Int. Conf. Wear of Materials* (San Francisco), 1981, p 548-563

16. Levy, A.V.; Aghazadeh, M.; and Hickey, G., The Effect of Test Variables on the Platelet Mechanism of Erosion, *Wear*, Vol 108 (No. 1), 1986, p 23-42

17. Bellman, R., Jr., and Levy, A.V., Erosion Mechanism in Ductile Metals, *Wear*, Vol 70 (No. 1), 1981, p 1-29

18. Ninham, A., and Levy, A.V., The Erosion of Carbide-Metal Composites, *Wear*, Vol 121 (No. 3), 1988, p 347-361

19. Kosel, T.H., and Aptekar, S.S., Effect of Hard, Second Phase Particles on the Erosion Resistance of Model Alloys, Paper No. 113, *Corrosion '86*, National Association of Corrosion Engineers (Houston), 1986

20. Ball, A., and Patterson, A.W., Microstructural Design of Erosion Resistant Hard Materials, *Proc. 11th Int. Plansee Seminar*, Vol 2, Metallwerk Plansee, Reutte, Austria, 1985, p 377-391

# Chapter 4
# Effects of Erodent Particle Characteristics on the Erosion of Steel

## Particle Composition and Shape Effects

Unlike other types of mechanical testing of materials, the control of the test conditions in small, solid-particle erosion has limits. Variations beyond the control of the tester can have significant effects on the behavior of the material. Similarly, unknown, or uncontrolled aspects of the local operating conditions of an eroding material in service compound the problem of being able to precisely determine in the laboratory the wastage rates of materials used in components serving in erosive environments. Extensive efforts have been made, essentially without success, to develop such things as an analytical description of particle shape or a fluid mechanics based prediction of particle flow in terms of local velocity, impact angle, and particle solids loading in the carrier fluid at the immediate surface of the material being eroded. Material wastage rates are significantly affected by these variables; yet they cannot be precisely described for any particular test or in-service circumstance. This situation is one of the reasons why analytical models of erosion cannot predict material wastage without using sufficient test-derived constants to make the models empirical expressions that only apply to the particular materials and test conditions used in deriving the constants.

Yet, much is known about the erosion of materials that can provide guidance to the equipment designer and operator to enable them to have lower surface loss rates and resulting longer life using the most economically attractive materials possible. In this chapter, the effects of particle characteristics (both static properties and dynamic movements) on material wastage will be presented, within the limits available to define, measure and control them.

In order to understand the nature of the erosivity of various kinds of particles that can occur singularly or together in particle-containing flows in fluidized-bed combustors (FBC), pulverized-coal boilers (PCB), and other types of chemical process systems, a series of experiments was carried out using several different particles (Ref 1). Six particle compositions were used separately to erode AISI 1020 plain carbon steel and the resulting erosion rates were compared. The possible reasons for the erosion rates measured were assessed, and the role of particle strength/integrity was postulated. In addition, the effect of particle shape on erosivity was investigated using angular steel grit and spherical steel shot. The literature has considerable erosion data on ductile metals obtained using a number of different erodents: $SiO_2$ (Ref 2), $Al_2O_3$ (Ref 3), SiC (Ref 4), steel shot and grit (Ref 5), and mixed groups of particles obtained from components operating in eroding environments (Ref 6). Understanding the erosivity of various erodents is helpful in comparing data of various investigators in the literature.

Cold-rolled AISI 1020 steel was eroded at 25 °C with particles of the six different compositions at a velocity of 80 m/s. The impact angles were 30 and 90°. The 30° angle was selected as it is near the angle at which maximum erosion occurs. The 90° angle was used as it is the angle where a significantly lower erosion rate occurs on ductile metals. The compositions of the erodent particles and their hardnesses and densities are given in

Table 4.1. Hardness was used as a means of rating the overall strength and integrity of the erodents. The particles of the five different brittle material compositions were all angular in shape and were in the size range 180 to 250 μm. The two different shapes of the same ductile steel particle composition used to study the effect of shape on erosion rate were spherical (shot) and angular (grit) with an average size of 100 μm.

### Particle Strength

The steady-state erosion rates of the 1020 steel (150 HV) eroded by each type of particle are

**Table 4.1  Erodent particles and rates of erosion of AISI 1020 steel**

| Particle composition | Density, (g/cm$^3$) | Mohs hardness | Vickers hardness | Erosion rates, $\times 10^{-4}$ g/g $\alpha = 30°$ | Erosion rates, $\times 10^{-4}$ g/g $\alpha = 90°$ |
|---|---|---|---|---|---|
| CaCO$_3$ (calcite) | ... | 3 | 115 | 0.03 | Not measurable |
| Ca$_5$(PO$_4$)$_3$ (apatite) | ... | 5 | 300 | 0.5 | 0.3 |
| SiO$_2$ (sand) | 2.7 | 7 | 700 | 3.0 | 1.6 |
| Al$_2$O$_3$ (alumina) | 4.0 | 9 | 1900 | 2.6 | 1.4 |
| SiC (silicon carbide) | 3.2 | >9 | 3000 | 3.3 | 1.9 |
| Steel grit | 7.9 | ... | ... | 5.3 | ... |
| Steel shot | 7.9 | ... | ... | 1.4 | ... |

For AISI 1020 steel, Vickers hardness (HV) = 150 kgf/mm$^2$

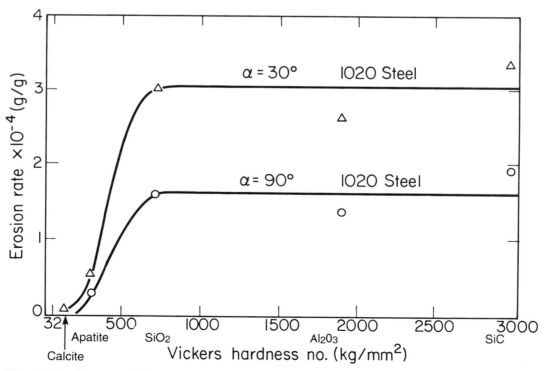

**Fig. 4.1**  Erosion rates of 1020 steel by calcite, apatite, SiO$_2$, Al$_2$O$_3$, and SiC

listed in Table 4.1. The rates for the five brittle erodents are plotted in Fig. 4.1. It can be seen that the erosion rates are very low for the softest, weakest erodent materials, calcite and apatite. Once the Vickers hardness of the particle reaches approximately 700 HV, indicative of particles strong enough not to shatter when they strike the target steel, the erosion rates remain essentially constant as the hardness/strength of the particles increases further. Thus $SiO_2$ at 700 HV has nearly the same erosivity as silicon carbide (SiC) at 3000 HV even though there is over four times difference in their hardness. The relatively small erosivity differences among $SiO_2$, $Al_2O_3$, and

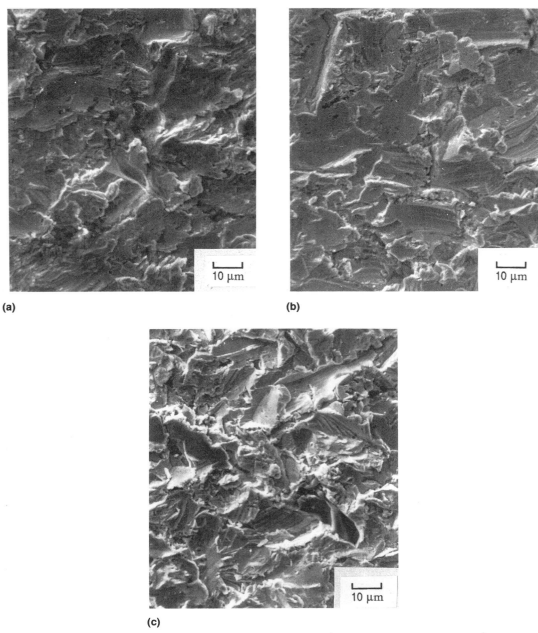

**(a)**

**(b)**

**(c)**

**Fig. 4.2** Surfaces of 1020 steel eroded by (a) $SiO_2$ ($E_r = 3.0 \times 10^{-4}$ g/g), (b) $Al_2O_3$ ($E_r = 2.6 \times 10^{-4}$ g/g), and (c) SiC. ($E_r = 3.3 \times 10^{-4}$ g/g), $\alpha = 30°$, $V = 80$ m/s

SiC are primarily due to small differences in the angularity of the particles, with SiC having the sharpest angles in the as-crushed powder. The strength/integrity of the erodent particles determines whether they can impact the target surface without shattering. This, in turn, establishes their size at impact and the resultant kinetic energy they impart to the target. Their kinetic energy is a primary determinant of their erosivity.

Figure 4.2 shows surfaces of 1020 steel eroded by $SiO_2$, $Al_2O_3$, and SiC. The appearance of the craters and platelets is very similar for all three erodents. The shape of the craters caused by

**Fig. 4.3** $Al_2O_3$ erodent particles (a) before impact, (b) after impact at 30°, and (c) after impact at 90°. $V = 80$ m/s

Al$_2$O$_3$ is a little less severe than those caused by SiC, which is thought to account for the slightly lower erosion rate caused by Al$_2$O$_3$. Figure 4.3 shows typical Al$_2$O$_3$ particles before and after impacting on the steel targets. The Al$_2$O$_3$ parti-

cles had a somewhat more rounded shape than the SiO$_2$ and SiC particles, but were generally similar. The size and shape of the three higher hardness erodents were the same after impact as they were before, indicating that they did not shatter upon

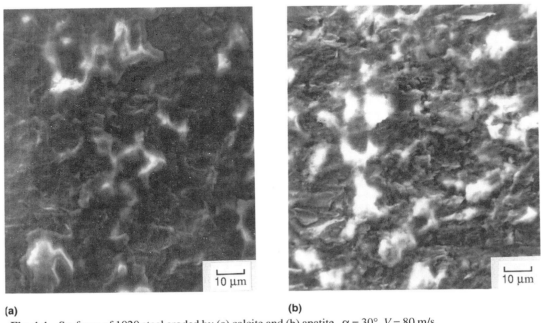

**(a)**  **(b)**

**Fig. 4.4** Surfaces of 1020 steel eroded by (a) calcite and (b) apatite. $\alpha = 30°$, $V = 80$ m/s

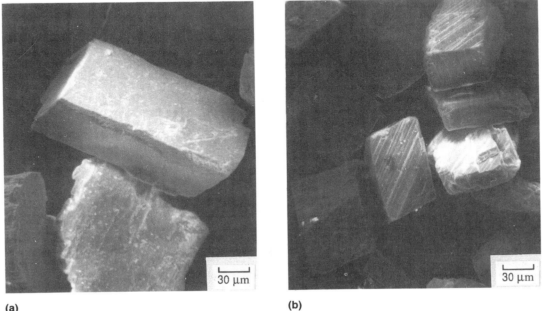

**(a)**  **(b)**

**Fig. 4.5** Calcite particles (a) before impact and (b) after impact on 1020 steel. $\alpha = 30°$, $V = 80$ m/s

impact. The striations seen in the shallow craters of eroded surfaces in Fig. 4.2 are imprints of the contour facets of the particles made as they translate along the crater surface. Figure 2.13 in Chapter 2 shows these facets on a SiC particle (Ref 7).

Figure 4.4 shows the eroded surfaces of the AISI 1020 steel impacted by calcite and apatite particles. Significant amounts of calcite and apatite are smeared on or embedded in the steel. The production of platelets and shallow craters is not as readily defined as in Fig. 4.2, although the platelet-formation mechanism appears to be the active mechanism on both surfaces shown. The breakup of the weak particles and their adherence to the eroded surface both decrease the kinetic energy of incoming particle segments and cover over the eroding surface with a protective layer of particle fragments. The calcite particles before and after impact are shown in Fig. 4.5. The marked difference in size due to fracture on impact is easily seen. The basic shape of the impacted particles appears to be similar to that of the particles before impact. Thus, it is the particles' kinetic energy, which is dependent on their size and density, that has the greatest effect on their erosivity, more so than their shape.

Incremental erosion rate curves resulting from erosion at 90° impact angle by the weak apatite particles and the strong $Al_2O_3$ particles give further insight into the effect of particle strength on erosion. Figures 3.4 and 3.5 in Chapter 3 show the erosion rate curves obtained using the apatite particles that broke up on impacting the AISI 1020 steel and the $Al_2O_3$ particles that did not. It took almost 200 g of apatite particles to reach steady-state erosion while the $Al_2O_3$ particles only took 50 g. The steady-state erosion rate of 1020 steel caused by apatite particles was less than $0.3 \times 10^{-4}$ g/g while that from the $Al_2O_3$ particles was $1.4 \times 10^{-4}$ g/g.

### Particle Shape

In order to determine how the shape of the particles affected their erosivity, same size steel particles in both angular and spherical shapes were used to erode the AISI 1020 steel. An impact angle of 30° at 25 °C was used. Figure 4.6 shows the particles after impacting on the target surface. No fracturing of the steel particles was observed, so their integrity was in the range of those of the $SiO_2$, $Al_2O_3$, and SiC particles.

(a)

(b)

**Fig. 4.6**   (a) Steel grit and (b) steel shot erodents after impact. $\alpha = 30°$, $V = 80$ m/s

Figure 4.7 shows surfaces of AISI 1020 steel eroded by the steel particles. The erosion rate $E_r$ caused by the angular steel grit was four times greater than that caused by the spherical steel shot, that is, $5.3 \times 10^{-4}$ g/g compared with $1.4 \times 10^{-4}$ g/g. The appearance of the eroded surfaces indicates the reason for the difference. The angular steel grit caused much sharper, deeper craters to form, which caused a more efficient production of extruded platelets. The spherical steel shot developed more shallow, rounded craters that did not produce platelets as efficiently. The production of platelets and the factors that influence their formation, extension, and removal are discussed below and in Chapter 2.

### *Crater, Platelet Formation*

The erosivity of impacting particles is primarily a function of the concentration of force that the particle can cause in a microscopic area of the target metal. When particles are so weak and friable that they cannot maintain their integrity when they strike the metal surface (calcite and apatite) they shatter into many smaller pieces. These pieces do not have the mass necessary to provide the localized force that can form platelets

efficiently and subsequently remove them by exceeding the local fracture stress of the target metal.

When erodent particles reach a sufficiently high level of strength and integrity where they do not fracture on impact for the velocity regime used, the erosion rate for particles of similar shape and density becomes approximately the same. Thus $SiO_2$, $Al_2O_3$, and SiC, which have similar shapes and are in the same density range, have about the same erosivity. Steel grit has a higher density and is considerably more erosive (Table 4.1). Thus, the kinetic energy of the particles when they strike the surface plays an important role in determining the amount of force that is available to make and deform platelets and to cause subsurface work hardening (Ref 7).

Local concentration of force is also a function of the geometry of particles. Angular particles can concentrate this force more effectively than rounded particles. This effect can be quite subtle. For a given target material with a fixed level of ductility such as the 1020 steel, the small difference in angularity between $Al_2O_3$ and SiC particles can be detected by the resulting erosion rates (Table 4.1). The great difference in shape be-

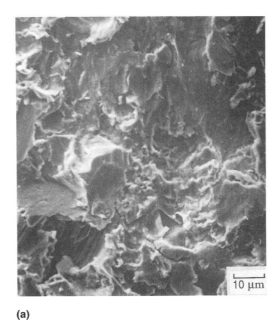 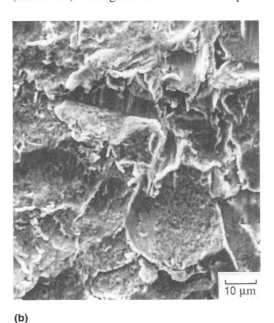

**(a)**          **(b)**

**Fig. 4.7**  Surfaces of 1020 steel eroded by (a) steel grit ($E_r = 5.3 \times 10^{-4}$ g/g) and (b) steel shot ($E_r = 1.4 \times 10^{-4}$ g/g) $\alpha = 30°$, $V = 80$ m/s

tween the steel grit and the steel shot causes a great difference in erosivity.

The kinetic energy of the particles also affects the time required to reach steady-state conditions. The amount of the strong $Al_2O_3$ particles that did not shatter required to reach steady-state erosion was considerably less than the weak apatite particles that shattered upon impact. Their larger mass and attendant larger kinetic energy could concentrate more force in the eroding metal and, hence, develop a steady-state number of platelets and the required subsurface cold-worked zone sooner (see Chapter 2). This effect may also relate to the particle size effect in erosion that has been reported extensively in the literature (Ref 8, 9). The lower erosion rate caused by the particles that shattered on impact is also partially the effect of the erodent particles that remain on the surface of the specimens and the protective layer that they develop.

## Particle Feed Rate and Size Effects(Ref 10)

### Spherical Erodent

The curves in Fig. 4.8 show that there was a peak mass loss that occurred at the 300 μm particle size when spherical glass beads were used to erode 1018 steel at $V = 60$ m/s. A much lower loss occurred at $V = 20$ m/s. Below this size, the kinetic energy of the particles appeared to be too low for them to be as effective in removing material as the 300 μm size beads even though there are more of the smaller beads to impact the target in a given weight of erodent. This might be explained in part by the fact that the number of particles actually striking the surface does not increase in the same way as the number of particles traveling toward the specimen due to a shielding effect provided by the rebounding par-

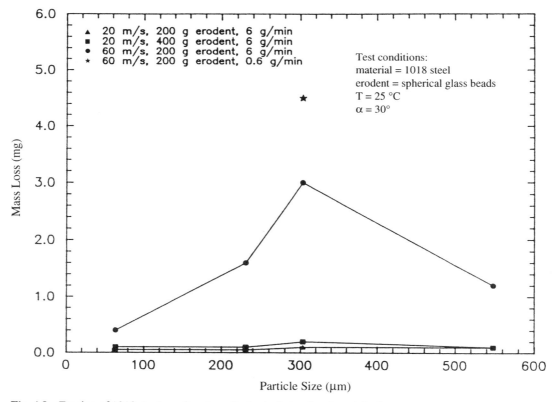

Fig. 4.8 Erosion of 1018 steel as a function of spherical glass bead particle size

ticles (Ref 8). Also, it seems that the decreasing kinetic energy of the smaller particles dominates the erosion behavior and the increasing number of particles in a given weight of erodent, called solids loading, does not result in higher mass loss as the particle size gets smaller.

At particle sizes greater than 300 μm, another factor became dominant in controlling the mass loss. The decrease in mass loss above this particle size was due in part to the decrease in the number of particles striking the surface. However, the major factor causing the reduction relates to the particle's ability to penetrate the target surface. While the larger diameter glass beads had more mass and, hence, more kinetic energy, the particle diameter became large enough to markedly decrease the ability of the spheres to actually penetrate the target surface and cause the severe plastic deformation that crater and platelet formation (Ref 7) require for effective removal of material.

At 60 m/s there was a considerably higher mass loss for the 300 μm glass beads that impacted the surface at the feed rate of 0.6 g/min compared to 6 g/min. This indicates that there was particle-to-particle interference at the higher solids loading that reduced the effectiveness of the particles to erode the surface. The primary mode of this interference was probably particles rebounding up from the surface deflecting incoming particles in the downward moving stream. Another example of this feed rate, or solids loading effect is shown in Fig. 4.9 for two stainless steels tested over a wide temperature range (Ref 11). The lower feed rate had a much higher erosion rate.

### Angular Erodent

Figure 4.10 shows the mass loss versus angular SiC particle size curve at a particle velocity of 20 m/s. The shape of the curve is the same as that determined by Tilly (Ref 9) for a much smaller range of particle sizes at a much higher particle velocity for a much stronger alloy. The steep slope of the curve at small particle diameters can be interpreted as the steep increase in particle kinetic energy with increasing particle size, as was found in the tests using spherical particles

(Fig. 4.8). The more or less constant mass loss with increasing particle diameter above 200 μm particle size is probably due to a combination of the relation between four characteristics of the particle stream that appear to influence mass loss:

- The particle size
- The number of particles striking the surface
- Their kinetic energy
- The interference between incoming and rebounding particles

Metallographic observation of particles of various sizes indicates that sharpness of the SiC particle edges does not change with increasing particle size and, therefore, there is no decrease in their ability to penetrate and plastically deform the metal surface as occurred with the spherical particles. Thus, the curve in Fig. 4.10 remains flat over a wide range of particle sizes. Note that only the largest particle loading (6.0 g/min) at the 250 μm angular particle size resulted in a decreased metal loss, as occurred with the 300 μm spherical

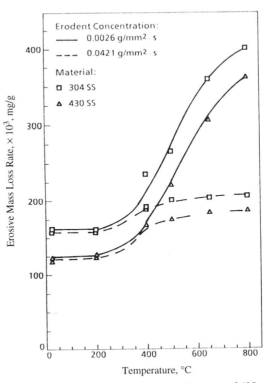

**Fig. 4.9** Variation of erosive mass loss rate of 430 and 304 stainless steels with temperature for two erodent concentrations

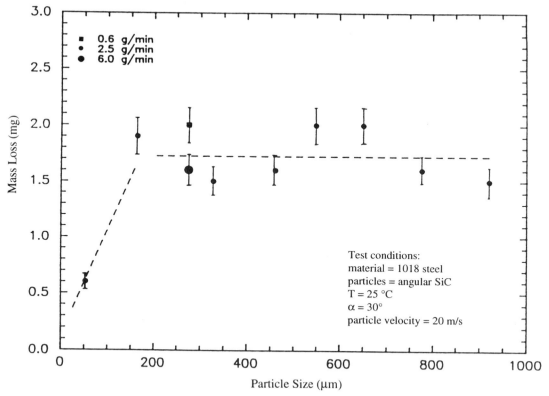

**Fig. 4.10**  Effect of particle size of angular SiC on metal wastage. $\alpha = 30$, $V = 20$ m/s

**Table 4.2   Effect of particle shape on erosion of 1018 steel**

| Particle size, μm | Feed rate, g/min | Mass loss, mg | | | |
|---|---|---|---|---|---|
| | | 20 m/s | | 60 m/s | |
| | | Spherical | Angular | Spherical | Angular |
| 250-355 | 6.0 | 0.2 | 1.6 | 3.0 | 28.0 |
| | 0.6 | 0.2 | 2.0 | 4.5 | 32.7 |
| 495-600 | 6.0 | 0.1 | ... | 1.2 | ... |
| | 2.5 | ... | 2.0 | ... | 42.4 |

particles, see Fig. 4.8. There was no difference in metal loss between the 2.5 and 0.6 g/min flow rates.

The results of the 60 m/s velocity tests are plotted in Fig. 4.11. A relatively steady increase of mass loss with increasing particle diameter occurred. The four particle stream characteristics mentioned earlier that are thought to influence the mass loss are combined in a manner that caused a nearly linear relation between particle size and mass loss. The change in slope in the range 150 to 300 μm may relate to the flattening of the curve that Tilly found (Ref 9). Why the marked change

in the shape of the curves for the 20 and 60 m/s velocities occurred is not known.

### Particle Shape

The effect of the shape of the erodent particles on the mass loss is shown in Table 4.2 for two size ranges of particles. It can be seen that the shape of particles is a major factor in establishing their erosivity (Ref 10). The difference is nearly a factor of 10 for the smaller size particles and up to almost 40 times for the larger size particles. While there is some difference in mass loss between a feed rate of 2.5 and 6 g/min (see Fig. 4.10

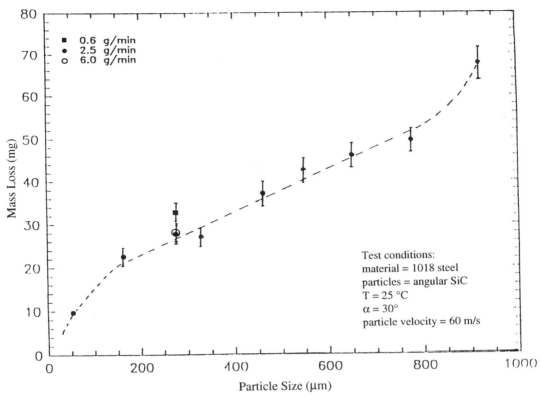

**Fig. 4.11**  Effect of particle size of angular SiC on metal wastage. $\alpha = 30$, $V = 60$ m/s

and 4.11), the last two lines of Table 4.2 can be compared. The shape factor is not significantly affected by differences in the feed rate, comparing 6.0 and 0.6 g/min rates for the 250 to 355 μm size particles, but there is the tendency for the lower flow rate to cause larger mass loss, as discussed below.

### Solids Loading

The results of the variation of solids loading, the weight of particles striking the surface per unit of time, or feed rate, for the 250 to 355 μm particles can be found in Fig. 4.8, 4.10, and 4.11. In general, the mass loss at the lowest solid loading was higher than at the higher solids loadings. The average distance between two particles moving toward the target surface, which depends on the particle size, the feed rate, and the density of the particle material, is listed in Table 4.3. As long as the distance between the particles is large enough, a low number of collisions between particles occurs. Most particles are able to strike the

surface and leave the area before the next particle strikes the same area. With decreasing distance between particles, more and more rebound particles collide with incoming particles and slow them down and/or change their trajectories. Both factors decrease the force or angle with which they strike the surface and, in some instances, prevent the incoming particles from even striking the target.

The effect of flow rate is shown for the 60 m/s spherical particles in Fig. 4.8. The difference in the distance between the particles for solids loadings of 0.6 and 6.0 g/min was one order of magnitude (Table 4.3). There was a 50% greater metal loss at the lower solids loading because the greater distance between particles prevented much of the interference between particles from occurring.

The results from the angular SiC particle tests plotted in Fig. 4.10 show that there was relatively little difference in mass loss as a function of

**Table 4.3    Average distance between particles (one dimension)**

| Particle size, μm | Feed rate, g/min | Velocity, m/s | Distance, mm |
|---|---|---|---|
| **Glass beads** | | | |
| 212-250 | 6 | 20 | 3 |
| 212-250 | 0.6 | 20 | 29 |
| 250-355 | 6 | 60 | 20 |
| 250-355 | 0.6 | 60 | 196 |
| **Silicon carbide** | | | |
| 250-300 | 6 | 20 | 7 |
| 250-300 | 2.5 | 20 | 17 |
| 250-300 | 0.6 | 20 | 70 |
| 250-300 | 6 | 60 | 21 |
| 250-300 | 2.5 | 60 | 50 |
| 250-300 | 0.6 | 60 | 210 |

loss for the smallest solids loading, in spite of the large difference in particle separation listed in Table 4.3. Why there is such a large difference between the solids loading effect for spherical particles and for angular particles can only be speculated. It probably is related to the greatly increased erosivity of the angular particles (Table 4.2), that reduces the effect of interference due to particle separation.

## References

1. Levy, A.V., and Chik, D., The Effect of Erodent Composition and Shape on the Erosion of Steel, *Wear*, Vol 89 (No. 2), 1983, p 151-162

2. Gulden M.W., Influence of Brittle to Ductile Transition on Solid Particle Erosion Behavior, *Proc. 5th Int. Conf. Erosion by Solid and Liquid Impact* (Cambridge), Sept 1979, Cavendish Laboratory, Cambridge University, 1979

3. Sargent, G., et al., The Erosion of Plain Carbon Steels by Ash Particles from a Coal Gasifier, *Proc. Int. Conf. Wear of Materials* (San Francisco), March 1981, American Society of Mechanical Engineers, 1981

4. Levy, A.V., The Solid Particle Erosion Behavior of Steel as a Function of Microstructure, *Wear*, Vol 68 (No. 3), 1981, p 269-287

5. Quadir, T., and Shewmon, P., Solid Particle Erosion Mechanisms in Copper and Two Copper Alloys, *Metall. Trans. A*, Vol 12, July 1981, p 1163-1176

6. Tabakoff W.; Hamed, A.; and Ramachandran J., Study of Metals Erosion in High Temperature Coal Gas Streams, *J. Eng. Power*, Vol 102 (No. 1), 1980, p 148-152

7. Bellman, R., Jr., and Levy, A.V., Erosion Mechanism in Ductile Metals, *Wear*, Vol 70 (No. 1), 1981, p 1-27

8. Misra, A., and Finnie, E.I., On the Size Effect in Abrasive and Erosion Wear, *Wear*, Vol 65 (No. 3), 1981, p 359-373

9. Tilly, G.P., A Two-Stage Mechanism of Ductile Erosion, *Wear*, Vol 23, 1973, p 87-96

10. Liebhard, M., and Levy, A.V., The Effects of Erodent Particle Characteristics on the Erosion of Metals, *Wear*, Vol 151 (No. 2), 1991, p 381-390

11. Zhou J. and Bahadur, S. "Further Investigations on the Elevated Temperature Erosion-Corrosion of Stainless Steels," Paper 13, Proc. NACE Conf. Corrosion-Erosion-Wear of Materials at Elevated Temperatures (Berkeley, CA), 1990

# Chapter 5

# Erosion and Erosion-Corrosion of Steels at Elevated Temperatures

## Elevated-Temperature Erosion of Steels

The use of structural steel alloys in the elevated-temperature combined erosion-corrosion environments of energy utilization and chemical process plant components has resulted in many instances of unacceptable levels of surface degradation. The straight erosion rates and mechanisms of several low-alloy and stainless steels used in process plants and boilers were determined at temperatures from room temperature to beyond their normal-use temperature using near inert gas atmospheres to prevent oxidation of the steels from occurring during the tests. Heat-treatable steels were tested in the range of their heat-treatment temperatures to determine whether changes in erosion behavior occurred when microstructural changes were occurring in the alloys.

Oxidation corrosion in combination with erosion was restricted so that the behavior of the metals themselves could be determined at elevated temperatures. In separate studies the contributions to overall surface degradation of static and dynamic elevated-temperature corrosion and combined erosion-corrosion were determined. In this way the role of each surface-loss mechanism in the combined attack that occurs in service environments could be better understood. Any synergistic behavior to either promote or retard metal loss was easier to identify and assess. Other investigators combined elevated-temperature erosion and corrosion in their work to simulate various service environments (Ref 1-3).

**Table 5.1  Composition of alloys eroded at elevated temperatures**

| Alloy | Composition, wt% | | | | | | | Heat treatment |
|---|---|---|---|---|---|---|---|---|
| | Cr | Ni | Mo | Mn | Si | C | Fe | |
| 1018 | ... | ... | ... | 0.8 | ... | 0.18 | bal | Annealed at 850-900 °C for 1 h |
| 2¼ Cr-1Mo | 2.25 | ... | 1.0 | 0.5 | 0.5 max | 0.15 | bal | Annealed at 900 °C |
| 5Cr-½Mo | 5.0 | ... | 0.5 | 0.5 | 0.5 max | 0.15 | bal | Annealed at 900 °C |
| 410 | 12.5 | ... | ... | 1.0 max | 1.0 max | 0.15 max | bal | Solution treated at 925 °C for 30 min; tempered at 250 °C, 425 °C, 750 °C |
| 304 | 18.0 | 8.0 | ... | 2.0 | 1.0 | 0.08 | bal | Annealed at 1020 °C |
| 310 | 25.0 | 20.0 | ... | 2.0 | 1.5 | 0.25 | bal | Annealed at 1100 °C |
| 17-4PH(a) | 16.5 | 4.0 | ... | 1.0 max | 1.0 max | 0.07 max | bal | Condition B; hardened at 500 °C |

(a) Copper, 3.6 wt%

In order to restrict the surface behavior occurring to erosion only, undried nitrogen was used to carry the erodent through the nozzle. Therefore, the test results obtained cannot be used directly to represent metal-loss rates in plant environments. Combined erosion-corrosion will be discussed later.

### Experimental Conditions

The alloys tested are listed in Table 5.1 along with representative compositions from the literature. They were selected to have a variety of metallurgical responses to elevated-temperature exposure and to have a varying chromium content to resist the partial pressure of oxygen in the undried nitrogen. The materials were obtained from flat sheet or from 2.5 cm thick pipe sections. In the latter case, the specimens were run through rolls to produce 0.3 cm thick flat pieces. The materials were fully annealed prior to testing or, in the case of stainless steel type 410 and 17-4PH, heat treated as indicated in Table 5.1. The specimens were 5 by 2 by 0.3 cm and were polished prior to testing to a 240 or 600 grit finish.

The room-temperature erosion tests were conducted using the nozzle tester described in Chapter 1. The laboratory elevated-temperature tests were carried out in the elevated-temperature tester shown in Fig. 5.1, except for the reactive gas tests described in the last section of this chapter. This nozzle tester is essentially the same as the room-temperature tester described at the beginning of Chapter 1 except that a furnace encapsulates a gas-particle mixing chamber above the nozzle tube, an air-heating coil around the nozzle tube, the nozzle tube itself, and the specimen chamber and particle exit area. The left-hand sketch shows the entire tester. The small hopper on the top holds the particles. The particles flow down a vibrating tube into the mixing chamber where they are mixed with incoming heated air which propels them down the 42 cm long by 0.5 cm nozzle and impacts them on the flat specimen that is located 1.25 cm below the nozzle.

All of the tester components shown on the sketch on the right side of Fig. 5.1 are within a resistance heated furnace that is controlled to ±5

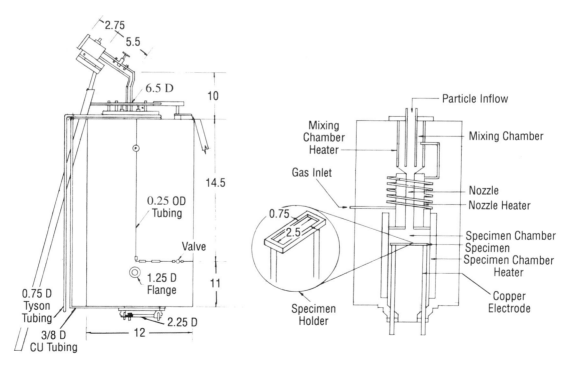

**Fig. 5.1**   Sketch of elevated-temperature nozzle test machine

°C. It can use a variety of erodents and either air, argon, or nitrogen carrier gases. Temperatures from 20 to 900 °C are achievable for the required time periods with a temperature variation not exceeding 15 °C over the test range.

The elevated-temperature erosion tests were generally conducted in one cycle using 300 g of 240 to 325 μm diam silicon carbide particles and 10 g/min particle loading. Some of the tests on the stainless steel type 410 were carried out incrementally. The specimen was preheated in the apparatus with a flow of nitrogen passing over it. The particles impinged upon the specimen surface at several velocities that were controlled by varying the pressure drop across the nozzle in a calibrated manner at impact angles of either 30° or 90°. The pressure setting was determined by using a computer program (Ref 4) which accounts for the elevated temperature of the gas-particle stream.

After completion of each experiment the specimen was quickly transferred to a cold-nitrogen flow and cooled to below 250 °C to prevent oxidation. Some oxidation of the specimens did occur, especially on low chromium content specimens at higher temperatures. The cooled specimens were weighed on an analytical balance accurate to 0.1 mg after being cleaned with alcohol in an ultrasonic cleaner.

### Austenitic Stainless Steels

Types 310 and 304 were tested at temperatures up to 900 °C. Figure 5.2 plots all of the test points measured for type 310. Four separate runs were made over the test-temperature range, using a different specimen for each test to determine the reproducibility of the tests. The locations of the data points indicates that the reproducibility is acceptable. Increasing the bulk temperature of alloys without introducing general oxidation corrosion can either increase or decrease the erosion rate of a particular alloy up to some intermediate temperature. Above this temperature, all metals undergo a relatively rapid increase in erosion rate with increasing temperature.

It can be seen that there is a marked difference in the amount of erosion and the shape of the erosion rate versus temperature curves between

the tests run at impact angles of 30° and 90°. The 30° curve shows that the erosion rate remained essentially the same until a test temperature near 400 °C was reached, at which point a rapidly increasing erosion rate with test temperature occurred. The 90° curve in Fig. 5.2 has a minimum occurring around the same 400 °C temperature, at which point the erosion rate began to rapidly increase as occurred in the 30° impact angle test. The absolute magnitude of the erosion is near the same at both angles near room temperature but differs markedly at higher temperatures. Erosion increased with particle velocity but was less sensitive to changes in velocity than was the case in room-temperature tests.

Figure 5.3 shows several micrographs of the eroded surfaces of 310 at three temperatures and two impact angles. It can be seen from the appearance of the surfaces that the mechanism of erosion at all temperatures and both angles is the

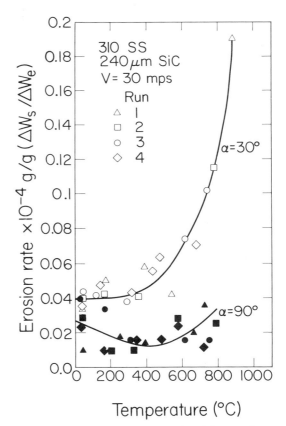

**Fig. 5.2** Erosion rate of type 310 stainless steel versus test temperature

**30° Impingement Angle**    **90° Impingement Angle**

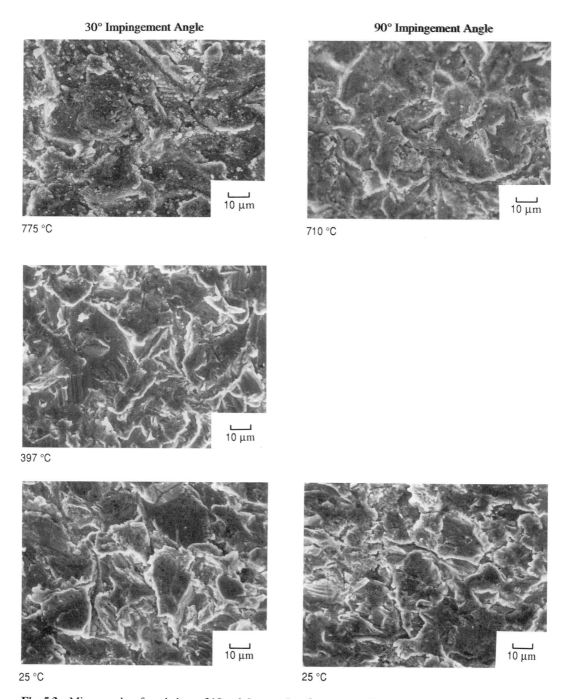

775 °C    710 °C

397 °C

25 °C    25 °C

**Fig. 5.3** Micrographs of eroded type 310 stainless steel surface at several test temperatures

same, even though the rates of erosion are different. The difference in the size of the shallow craters and platelets appears to increase somewhat with test temperature, but not with impact angle. Both impact angles result in eroded surfaces that look the same, even though the actual amount of erosion is less at 90° than at 30°. These observations verify other work that the basic

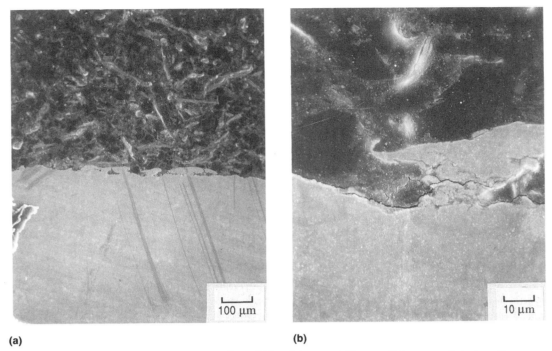

**(a)**            **(b)**

**Fig. 5.4**   Cross sections of eroded surface of type 310 stainless steel. Test conditions: nozzle tester, nitrogen gas, $\alpha = 30°$, $T = 710 \,°C$

mechanism of erosion of ductile metals does not change with impact angle (Ref 5).

Figure 5.4 shows a cross section of the surface area of an eroded 310 specimen at two different magnifications. The platelets and craters that occur on the eroding surface can be seen. At the higher magnification (Fig. 5.4b), the stem of the platelet (located at the center of the eroded surface in Fig. 5.4a) attaching it to the base metal is in a severely fractured condition, indicating that the platelet will probably be removed by the next particle that strikes it.

Figure 5.5 shows the curve of steady-state erosion rate versus test temperature for 304 at a 30° impact angle. This steel has a lower chromium content (18%) than 310 (25%). The shape of both the 304 and 310 curves are the same, but the erosion rates were significantly different above 400 °C, where the erosion rate of 304 is three times the rate of 310. While both alloys were eroded in an undried nitrogen gas atmosphere to reduce oxidation, there was sufficient oxygen present to cause higher loss rate, com-

bined erosion-corrosion to occur in the lower chromium content steel.

### Low-Alloy Steels

Three different steels with increasing chromium contents from 0 to 5% Cr were tested at 30° impact angle. The curves of erosion rate versus test temperature for the materials are plotted in Fig. 5.6. It can be seen that they behave in a similar manner. Their curves slope down somewhat from room temperature to approximately 200 °C and then turn and begin to show an increasing erosion rate with test temperatures starting at 250 to 300 °C. It is postulated that the initial decrease in the erosion rate was due to an increase in the overall ductility of the bulk of the test specimen and a decrease in its subsurface work hardening. While the short-time tensile elongation remained level or even decreased somewhat for some of the alloys up to 400 °C, the impact strength of some of the alloys increased with temperature. This distributed the force of the impacting particles by plastic deformation of the subsurface region enough to reduce the magni-

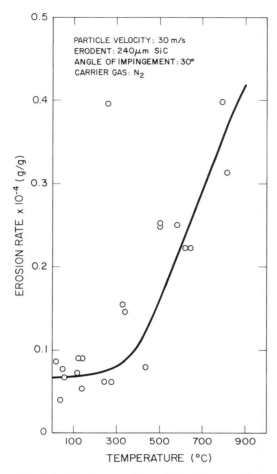

**Fig. 5.5** Erosion rate of type 304 stainless steel versus test temperature

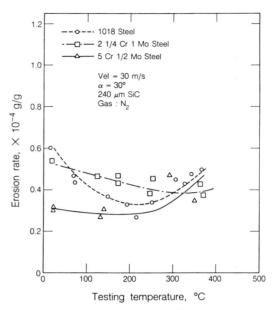

**Fig. 5.6** Erosion rates of 1018, 2¼Cr-1Mo, 5Cr-½Mo steels versus test temperature

tude of the localized stresses that developed in the immediate vicinity of each particle's impact zone. This reduced the amount of extrusion, forging, and fracture of platelets at the eroding surface and, hence, the erosion rate. This is the same type of pattern that was determined for the austenitic stainless steels, except that for 304 and 310, the 30° angle curves did not dip prior to the temperature at which the erosion rates increased with test temperature.

The eroded surfaces of the steels tested all showed that the platelet mechanism of erosion had occurred. Figure 5.7 shows the 2¼Cr-1Mo steel and 310 stainless steel surfaces after erosion at a test temperature of near 400 °C. There is an indication that oxidation is beginning to occur on

the tested surface of the low chromium content steel, as indicated by a speckled appearance that develops, even though the specimens are bathed in a nitrogen flow before, during, and after each test until they have cooled sufficiently below their oxide-forming temperature. The 310 specimen with 25% Cr shows no evidence of oxidation of the test surface.

A fairly common characteristic observed in the erosion of many ductile metals above the temperature where their erosion rates start to increase with test temperature is the formation of narrow, elongated gouges, as can be seen in Fig. 5.7(a). These occurred on all but the stainless steels. They are formed by sharp, angular protrusions that occur at various angles on the erodent particles. They penetrate the soft surface of the steel and are moved along for a distance by the momentum of the particle. These marks are generally not found in steels tested at room temperature or in austenitic stainless steels. The presence or absence of the gouges may be a function of the number of active slip planes, which are different for bcc ferritic steels and fcc austenitic steels.

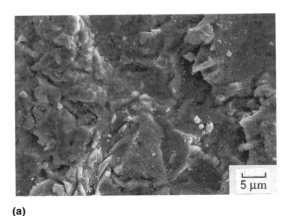

**(a)**

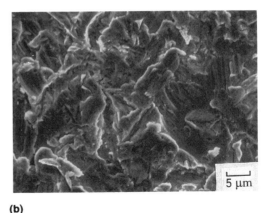

**(b)**

**Fig. 5.7** Eroded surfaces of (a) 2¼Cr-1Mo steel and (b) type 310 stainless steel. Test conditions: nozzle tester, nitrogen gas, 300 g of 240 μm SiC, velocity = 30 m/s, α = 30°, temperature: (a) 370 °C and (b) 397 °C

## Heat Treatment Hardenable Stainless Steels

Type 410, a martensitic-hardening steel, and 17-4PH, a precipitation-hardening steel, were tested to determine how they respond when erosion tested in the region of their heat-treatment temperatures. Also, the brittle behavior of type 410 when tempered at 475 °C was studied by erosion testing at temperatures below, at, and above its temper brittle inducing temperature. Type 410 was also used to determine the precise shape of the incremental erosion rate curve when small erodent increments were used.

The tempering temperature and, particularly the temper brittle tempering temperature had no effect on the elevated temperature erosion rate of 410. Figure 5.8 plots the results of all three tempers on one curve of erosion rate versus test temperature. There were no breaks in the curve to account for the test being carried out at a particular tempering temperature. The shape of the curve is the same one that occurred for type 310, 1018, and 5Cr-½Mo steels. The reduction in the rate of erosion at the intermediate test temperatures was greater than occurred for the other steels.

Precipitation-hardened 17-4PH stainless steel in the heat-treated condition behaved in the same manner as most of the other steels tested, having a minimum erosion rate at intermediate test temperatures in a 30° impact test. Figure 5.9 shows that testing in the range of the 500 °C heat-treatment temperature did not cause any breaks to

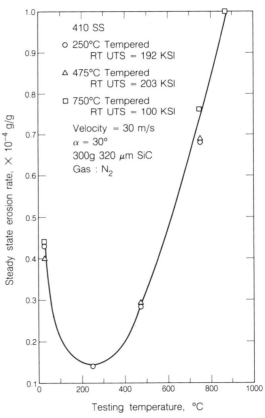

**Fig. 5.8** Erosion rate of type 410 stainless steel versus test temperature

occur in the curve. There does not appear to be any readily discernible effect of testing at the heat-treatment temperatures on the erosion rates of the two heat-treated alloys.

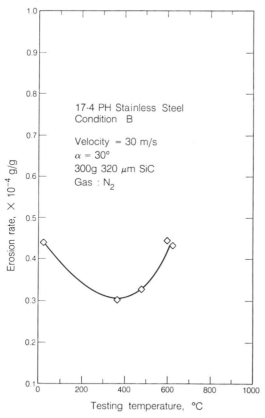

17-4 PH Stainless Steel
Condition  B

Velocity  = 30 m/s
$\alpha = 30°$
300g 320 $\mu$m SiC
Gas : $N_2$

**Fig. 5.9** Erosion rate of 17-4PH stainless steel versus test temperature

The bar graphs in Fig. 5.10 show how the steady-state erosion rates of all of the materials tested compared at 250 °C. Austenitic steels had considerably lower erosion rates than the ferritic steels. The austenitic steels were tested in their most ductile, fully annealed condition, while the ferritic steels were tested in a hardened, lower ductility condition. The ductility differences probably accounted for the erosion-rate differences. Heat treatment hardness levels had no relation to erosion rates (Ref 6). The erosion rates for the ferritic 410 were in the range of those of the austenitic steels.

## Combined Erosion-Corrosion of Chromium Steels

The ability of alloy steels to withstand the surface degradation caused by the combined ero-sion-corrosion surface environments that occur in such energy-production systems as fluidized-bed combustors, advanced pulverized coal boil-ers, and entrained-coal gasifiers is an important aspect of their design. Considerable effort has been expended to gain an understanding of the elevated-temperature corrosion that occurs (Ref 7-12). The effects that simultaneous erosion and corrosion have on the behavior of the steels com-monly used in these systems are different in terms of wastage rates and mechanisms from either corrosion or erosion, separately. The two degra-dation processes are not necessarily synergistic. Sometimes the formation of a protective surface layer occurs, formed in part by the oxide from the base metal. In many reported instances of metal loss, corrosion scale on the surface of the exposed alloys has been interpreted as the absence of an erosion component in the metal-wastage process. This is not the case. The scale material at the exposed surface is always eroding at some rate, and this loss, in turn, does result in base-metal loss.

The impact of small, solid, erodent particles on the surface of steels at elevated-temperature conditions where oxidation occurs markedly changes the mechanisms and rates of surface degradation that occurred at the near inert gas conditions described earlier. The changes vary with the surface environment, that is, gas compo-sition and temperature and particle composition, shape, integrity, velocity, impact angle, and con-centration. Their effects on the corrosion scale that forms changes its composition, morphology, growth and loss mechanisms, and the resultant rates of metal wastage, compared to static or dynamic gas-formed corrosion scales.

Comparisons between the behavior of the same commercially available steels exposed to erosion-corrosion environments in laboratory test devices and in atmospheric fluidized bed combustors (AFBC) indicate that the mecha-nisms and the rates of metal wastage are very similar. Thus, the information developed under laboratory conditions on the combined erosion-corrosion of alloys can be used to better under-stand and even predict their in-service behavior. It is particularly important to know how impact-

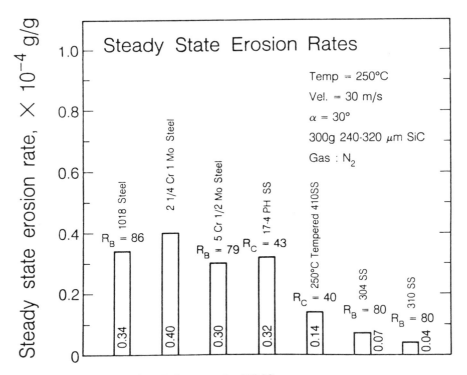

**Fig. 5.10** Erosion rate bar graph for all alloys tested at 250 °C

ing particles change the nature and behavior of the scales that form on the alloy surfaces and how the resulting scale loss translates to substrate metal loss. Without this understanding, the presence of scale on the surface of a tube in an AFBC could be misinterpreted as meaning that only corrosion was occurring and that the impacting particles had little or no erosive effect on metal loss.

The erosion-corrosion behavior of the steels changes as a function of their chromium content as well as the result of differences in the surface environment. The selection of the lowest-cost alloy for a required service is aided by an in-depth understanding of erosion-corrosion behavior. Use of imprecise, synergistic effects of combining erosion and corrosion on metal wastage (Ref 13) as the basis for selecting materials are rejected and the actual mechanisms of behavior are presented in this chapter that can form the basis for more intelligent selection of alloys.

**Test Conditions**

**Laboratory Nozzle Tester.** All of the laboratory experiments except those performed in reactive gases were carried out in the elevated temperature nozzle tester shown in Fig. 5.1. Test times of 30 min, 2 and 5 h were used. The erosion tests of the steels were carried out at impact angles of 30° and 90°. The erodent particles were round-shaped $Al_2O_3$ and angular-shaped silicon carbide with an average particle size of 130 μm. Test temperatures somewhat above the normal service temperatures of some of the steels were used to obtain measurable surface losses in reasonable laboratory test durations. Duplicate or triplicate tests were run at each test condition.

Thickness changes of the base metal as the result of combined erosion-corrosion were made by measuring a cross section of the metal after testing using an optical micrometer to observe the cross section through the primary erosion zone.

The back side of each specimen was protected from the environment by setting it on a solid-surface cradle. Even so, a temper color thin oxide scale that was not measurable did occur on the back, protected side of the specimen. Weight loss measurements were made using a balance that indicated to 0.1 mg.

The steel specimens were cut from 1 in. thick, 5 in. inside diameter pipe. Their compositions are listed in Table 5.2. The specimens were flattened in rolls and annealed at 925 °C. Specimens with final size of 17.5 by 17.5 by 2 mm were milled and polished to a 600 grit finish.

To prevent oxidation of the test surface prior to the test, undried nitrogen was passed through the nozzle tester until the specimen reached the test temperature. When the transport of the erodent particles to the target surface began, the gas was changed to air. At the end of the erodent flow period, the air was stopped and nitrogen was again introduced into the system. After the test, the specimen was quickly removed from the fur-

**Table 5.2  Composition of eroded-corroded chromium steel alloys**

| Alloy | C | Fe | Cr | Ni | Mo | Others |
|---|---|---|---|---|---|---|
| 2¼Cr-1Mo | 0.15 | bal | 2.00-2.50 | ... | 0.90-1.10 | Mn 0.3-0.6; Si ≤ 0.5; S, P ≤ 0.03 |
| 5Cr-½Mo | 0.15 | bal | 4.00-6.00 | ... | 0.40-0.60 | Mn ≤ 1.0; Si ≤ 1.0; P < 0.04; S ≤ 0.03 |
| 9Cr-1Mo | 0.15 | bal | 8.00-10.00 | ... | 0.90-1.10 | Mn 0.3-0.6; Si 0.5-1.0; P, S ≤ 0.03 |
| Type 410 | 0.15 | bal | 11.50-13.50 | ... | ... | Mn, Si ≤ 1.0; P ≤ 0.045; S ≤ 0.03 |
| Type 304 | 0.08 | bal | 18.00-20.00 | 8.00-12.00 | ... | Mn ≤ 2.0; Si ≤ 1.0; P ≤ 0.045; S ≤ 0.03 |
| Type 310 | 0.25 | bal | 24.00-26.00 | 19.00-22.00 | ... | Mn ≤ 2.0; Si ≤ 1.0; P ≤ 0.045; S ≤ 0.03 |

(a)

(b)

(c)

**Fig. 5.11**   Scale morphology of 9Cr-1Mo steel at (a) and (b) 750 and (c) 900 °C. Test conditions: nozzle tester, air, $t$ = 30 min, $\alpha$ = 90°. Dynamic corrosion (a) and (c), $V_{air}$ = 60 m/s. Erosion-corrosion (b), $V_{particle}$ = 70 m/s

nace section of the tester and placed under a protective flow of nitrogen until it had cooled to approximately 300 °C to prevent further oxidation. Sometimes a small amount of spalling of the scale on the test surface occurred during cooling. An optical and a scanning electron microscope were used to observe the specimens' surfaces and cross sections. Energy dispersive spectroscopy (EDS) and x-ray diffraction (XRD) were used to determine the compositions of the scales.

**Fluidized-Bed Combustor.** The atmospheric fluidized-bed combustor (AFBC) that was used for the in-service tube exposures was located at the Point Tupper, Nova Scotia Power Corp. in Canada. Its operational characteristics and the

tube alloys that were exposed in the AFBC are discussed in Ref 14.

### Effect of Presence of Impacting Particles

The impact of small, solid particles at low velocities on a steel surface that is simultaneously undergoing elevated-temperature oxidation markedly changes the nature of the scale layer that is formed (Ref 15). The composition, morphology, growth rate, and thickness of the layer are all affected as is its removal mechanism and rate from the surface. The predominant feature of low particle velocity erosion-corrosion is the presence of an eroding scale layer on the surface of the metal rather than bare metal.

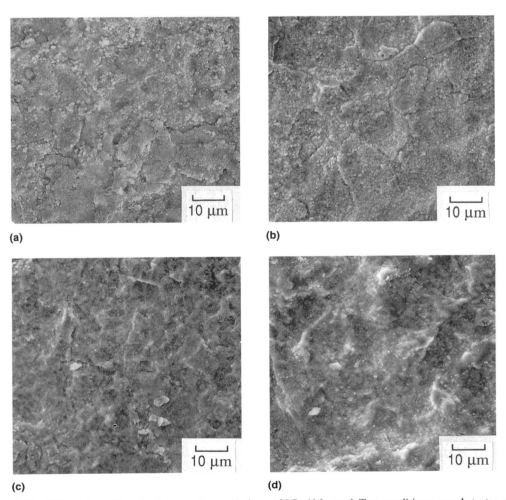

(a)　(b)

(c)　(d)

**Fig. 5.12** Effect of particle velocity on scale morphology of 9Cr-1Mo steel. Test conditions: nozzle tester, air, 130 μm Al$_2$O$_3$, α = 90°, $T$ = 850 °C, $t$ = 5 h, $V$ = (a) 10 m/s. (b) 30 m/s. (c) 45 m/s. (d) 70 m/s

A major difference in the formation mechanisms of scales, their morphologies, mechanical behavior, and loss rates was observed as a function of the impact angle of the erodent particles. A steep impact angle (90°) has a markedly different effect on the surface than a relatively shallow angle (30°). The mechanisms of loss that occurred at each angle are discussed separately. They form the basis for understanding the erosion-corrosion behavior of steels discussed in subsequent chapters.

The ability of the impacting erodent particles to increase the effective corrosion temperature of 9Cr-1Mo steel at a 90° impact angle is indicated by the morphology of the scale that forms (Fig. 5.11). Figure 5.11(a) shows the thin, adherent, essentially $Cr_2O_3$ scale that formed on the 9Cr-1Mo steel under dynamic corrosion conditions. In this test, air heated to 750 °C was directed at the surface at 60 m/s velocity with no erodent particles in it. When 130 µm $Al_2O_3$ particles were added to the 750 °C air, the scale morphology shown in Fig. 5.11(b) occurred. The scale consisted of domains of $Fe_2O_3$. The dynamic corrosion (no particles in gas flow) surface of the steel

at a test temperature of 900 °C is shown in Fig. 5.11(c). It is comparable in its morphology to that of the erosion-corrosion surface at 750 °C. Thus, the presence of erodent particles effectively increased the iron oxide scale forming conditions by 150 °C.

### 90° Impact Angle Tests

**Effect of Particle Velocity.** Increasing the velocity of the agglomerated, round-shaped $Al_2O_3$ erodent particles caused distinct changes to occur in the morphology of the scale surface at an impact angle of 90°. Figure 5.12 shows how the surface of the scale changed over a range of particle velocities after a 5 h exposure. The most notable difference occurred between a particle velocity of 30 and 45 m/s. At $V = 10$ and 30 m/s, the scale appeared to be segmented and cracked and had a larger oxide crystallite size (Fig. 5.12a and b). At $V = 30$ m/s, the beginning of consolidation of the scale can be seen. It became more pronounced at $V = 45$ and 70 m/s (Fig. 5.12c and d). At 45 and 70 m/s, the scale appeared to be essentially continuous, no division or cracks were discernible, and individual crystallites could not

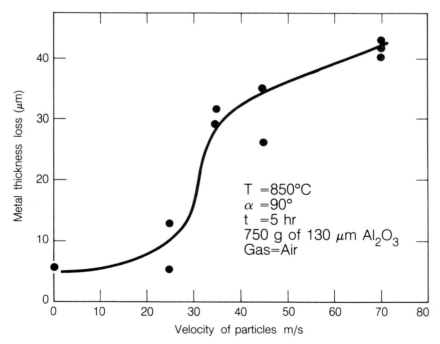

**Fig. 5.13** Effect of particle velocity on erosion-corrosion rate of 9Cr-1Mo steel

be observed at the magnification used. At 70 m/s, considerable amounts of embedded $Al_2O_3$ could be seen, primarily along the crests of the indentations.

The metal thickness loss versus particle velocity curve (Fig. 5.13) had a distinct transition in it at about 30 m/s going from low metal loss rates to high metal loss rates. The data point at 0 velocity is for dynamic corrosion, that is, with no particles in the flowing gas. This curve shape is indicative of a major change in the mechanism of the behavior that is being plotted. The major change in the mechanism of erosion-corrosion that occurred was the change in scale loss from a low loss rate mechanism of cracking and chipping of small pieces of the scale to a high loss rate mechanism of periodic spalling of larger pieces of consolidated scale. The loss of the scale at different rates translated to the loss of the base metal as it was oxidized to replace the eroding scale.

The spalling is thought to be due to the continuous nature of the scale formed at the higher particle velocities that did not have stress-reducing cracks in it as occurred in the scales formed at the lower-particle velocities. In the application of protective ceramic coatings on metals for gas-turbine components, cracks are induced in the coating during deposition to increase their thermal fatigue life and prevent spalling (Ref 16, 17). The large number of subcritical microcracks present reduces the elastic modulus and distance across continuous segments that minimize the stresses that can develop in the coating layer for a given strain level. This results in a strain accommodation that reduces the spalling tendency of the coating. This same effect is thought to govern the behavior of the scale on the 9Cr-1Mo steel under erosion-corrosion conditions. A small amount of spalling of the scale also occurred during the cooling down period from the 850 °C test temperature.

The change in metal thickness loss rate at the higher-particle velocities is thought to be due to the change in the mechanism by which the scale is removed, as described above, rather than because of a change from primarily corrosion to an undefinable, synergistic, combined erosion-corrosion mechanism. The initiation of loss of larger pieces of scale by spalling rather than by cracking and chipping could account for the increased metal loss. Such a sharp increase in erosivity of the particles could not occur between 30 and 40 m/s due only to increased particle velocity, because the basic erosivity of a material is a function of the kinetic energy of the impacting particles, which increases uniformly by the square of the velocity.

The surface areas shown in Fig. 5.12 for the higher velocity tests were in locations where spalling had not yet occurred in order to show the continuous nature of the scale. Figure 5.14 shows an area where both adhered and spalled scale were present. There is a pattern on the

(a)

(b)

**Fig. 5.14** Surface of 9Cr-1Mo steel showing adhered and spalled scale. Test conditions: nozzle tester, air, 130 μm $Al_2O_3$, $\alpha = 90°$, $V = 35$ m/s, $T = 850$ °C, $t = 5$ h

spalled area that is seen in the higher magnification micrograph to be a grain pattern in the scale that can be related to the base metal's grain size and shape.

Evidence that the principal spalling took place prior to the end of the test is shown in Fig. 5.15. Figure 5.15(a) shows the scale that formed after the spalling occurred. Figure 5.15(b) shows a scale area in the outer region of the specimen where no erodent particle impacts had occurred. The oxide grain size in the outer area is larger than that which formed in the spalled area because the outer area grains had a longer time to grow. They were exposed to the air atmosphere for the complete duration of the test. Also, it has been observed in other elevated-temperature erosion-corrosion testing (Ref 15), that erosion-corrosion enhances the growth of oxide-scale crystals. Therefore, the only way that the scale in the primary zone could have a smaller grain size than that in the outer zone is if it had considerably less time in which to grow.

**Effect of Test Temperature.** The effect of the test temperature on the metal thickness loss at the highest particle velocity (70 m/s) is shown in Fig. 5.16. The increase in the thickness loss between 750 and 850 °C was five times greater than that which occurred between 850 and 900 °C. This indicates that the loss mechanism had changed, as occurred as a function of velocity. That change was due to a change in the nature of the scale on the metal surface. As will be shown later, the same change occurred as a function of temperature as occurred as a function of particle velocity. Figure 5.16 is the transition portion of a classical S-shaped curve.

The effects of the test temperature and particle velocity on the morphology of the scale surface formed are shown in Fig. 5.17 and 5.18. The mechanism of scale loss as a function of both velocity and test temperature can be seen in this series of micrographs. Below the loss mechanism transition temperature or velocity, the scale is removed from the surface by the impacting erodent particles chipping away small pieces of scale that are segmented primarily as the result of its growth pattern. Above the transition temperature or velocity, the loss is by periodic spalling of larger pieces of scale as the result of stresses that build up in the more continuous, dense scale. The scale was condensed to a continuous layer by the hot pressing action of the impacting particles on the softened scale. It has been determined (Ref 18, 19) that the transition velocity is approximately 30 m/s and the transition temperature is approximately 750 °C for the scales formed on the 9Cr-1Mo steel used in this investigation.

The specimens tested at 850 °C show a change in scale appearance between the $V = 10$ m/s and $V = 70$ m/s tests (Fig. 5.17b and 5.18b, respectively). In the 10 m/s test (Fig. 5.17b), the scale is still segmented while in the 70 m/s test (Fig. 5.18b), the scale is consolidated and continuous with depressions occurring periodically in the

(a)

(b)

**Fig. 5.15** Morphology of scale beneath primary zone spalled area (a) and outer zone (b) of 9Cr-1Mo steel. Test conditions: nozzle tester, air, 130 μm Al$_2$O$_3$, $\alpha = 90°$, $V = 25$ m/s, $T = 850$ °C, $t = 5$ h

ductile scale as the result of particle impacts. The scale morphology change resulted in a scale loss mechanism change that is shown as a function of

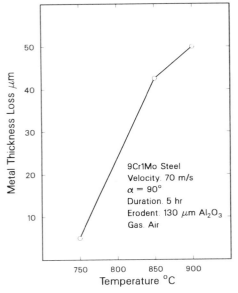

**Fig. 5.16** Metal thickness loss versus test temperature

temperature in Fig. 5.16 and as a function of velocity in Fig. 5.13. The segmented scales are removed from the surface at lower rates than the consolidated scales.

At 900 °C, the scale formed in the $V = 10$ m/s test (Fig. 5.17a) is still segmented and undergoes scale removal by a chipping mechanism. The scale in the 70 m/s test (Fig. 5.18a) is consolidated and has its scale removed by the periodic spalling mechanism. The differences in the continuity of the scales formed at different test temperatures can be seen most clearly in Fig. 5.18(a). Comparing the 900 °C, $V = 70$ m/s specimen surface in Fig. 5.18(a) with the 900 °C dynamic corrosion specimen surface in Fig. 5.11(c), shows how much scale consolidation is achieved by the hot-pressing type of action of the impacting particles.

**Effect of Carrier Gas.** The use of an inert gas, nitrogen, as the carrier gas for the erodent particles markedly reduced, but did not eliminate, the oxidation process. The nitrogen used in the tests was not dried and, therefore, had an oxygen con-

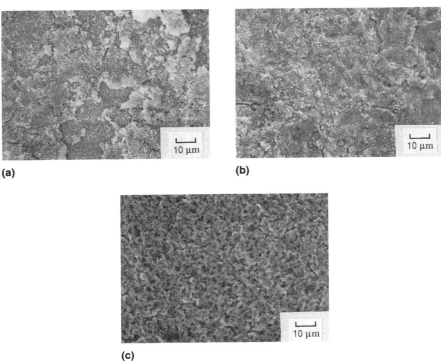

**Fig. 5.17** Effect of test temperature on scale morphology in $V = 10$ m/s test on 9Cr-1Mo steel. (a) $T = 900$ °C. (b) $T = 850$ °C. (c) $T = 750$ °C. Test conditions: nozzle tester, air, 130 μm Al$_2$O$_3$, $\alpha = 90°$, $t = 30$ min

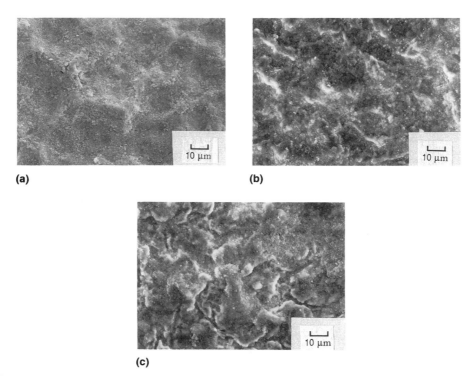

**Fig. 5.18**  Effect of test temperature on scale morphology in $V = 70$ m/s test on 9Cr-1Mo steel. (a) $T = 900$ °C. (b) $T = 850$ °C. (c) $T = 750$ °C. Test conditions: nozzle tester, air, 130 m $Al_2O_3$, $\alpha = 90°$, $t = 30$ min

tent. The morphology of the scale formed at 850 °C and 70 m/s particle velocity in a nitrogen-$Al_2O_3$ particle test is compared to the scale morphologies formed in air-$Al_2O_3$ particles tests at 650 and 850 °C in Fig. 5.19. The type and small domain size of scale formed at 650 °C in air (Fig. 5.19a) is similar to that formed in nitrogen at 850 °C (Fig. 5.19c). At 850 °C in air, the more continuous type of scale forms (Fig. 5.19b). The use of nitrogen as the carrier gas effectively reduced the erosion-corrosion temperature by 200 °C, which kept the scale-loss mechanism to the low-loss, chipping type. The large increase in domain size between the specimens tested in air at 650 and 850 °C at $V = 70$ m/s can be seen by comparing Fig. 5.19(a) and (b). Again, it is thought that it is this difference in scale continuity that is primarily responsible for the change in the mechanism of scale loss discussed earlier.

**X-Ray Diffraction Analysis.** Table 5.3 shows the results of the x-ray diffraction analyses performed on the various scales that formed on 9Cr-1Mo steel. The thickness of the scale was not

sufficient at temperatures below 650 °C to use x-ray diffraction. At 650 °C and above, it can be seen that $Fe_2O_3$ was the predominant scale. At 650 °C, it was the only constituent of the scale. Some fine, comminuted, submicron size $Al_2O_3$ erodent remained in the areas between the domains of the $Fe_2O_3$ scale after ultrasonic cleaning of the specimens prior to their analysis. At 850 and 900 °C, the scale was of a duplex type, as was observed in Ref 12 and 13. At 850 °C, the chromium-containing scale layer consisted of the spinel $FeCr_2O_4$, while at the higher temperature (900 °C) it consisted of a solid solution of $Cr_2O_3$ in $Fe_2O_3$.

### 30° Impact Angle Tests

**Analysis of Surfaces.** The scale formations that occurred on the 9Cr-1Mo steel due to combined erosion-corrosion at an impact angle of 30° using the same agglomerated, round shaped $Al_2O_3$ erodent had some similarities to those that formed at 90°, but a basically different kind of behavior occurred. For comparison purposes,

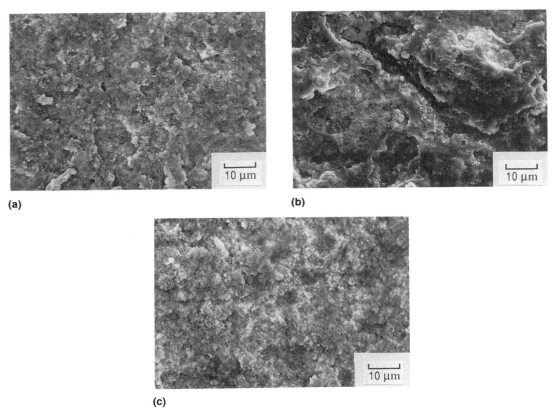

**Fig. 5.19**  Effect of air (a) and (b) and nitrogen (c) carrier gases on scale morphology of 9Cr-1Mo steel. Test conditions: nozzle tester, 130 μm Al$_2$O$_3$, α = 90°, V = 70 m/s, t = 5 h. Temperature: (a) 650 °C, (b) and (c) 850 °C

**Table 5.3    Effect of temperature on scale composition of 9Cr-1Mo steel**

| Test condition | X-ray diffraction analysis | | |
|---|---|---|---|
| 650 °C, 70 m/s, 5 h, 90° | α-Fe$_2$O$_3$ | ... | α-Al$_2$O$_3$ |
| 850 °C, 45 m/s, 5 h, 90° | α-Fe$_2$O$_3$ | FeCr$_2$O$_4$ | α-Al$_2$O$_3$ |
| 900 °C, 30 m/s, 30 min, 90° | α-Fe$_2$O$_3$ | Cr$_x$O$_y$ | ... |
| 900 °C, 45 m/s, 30 min, 90° | α-Fe$_2$O$_3$ | Cr$_x$O$_y$ | ... |

Fig. 5.20 shows the three different surface morphologies that occurred on a single specimen tested at an impact angle of 90° and a velocity of V = 35 m/s. The eroded footprint on the specimen consists of three distinct areas that can be visually discerned on the surface (Ref 20). The center of the eroded area is designated the primary zone. A peripheral area surrounding it is called the halo zone. An outer area where relatively few particles impact is the outer zone. It can be seen that scales of different composition and morphology occurred in the primary zone (Fig. 20a), the halo zone (Fig. 20b), and the outer zone (Fig. 20c).

Figure 5.21 shows a macroscopic view (a) of a specimen eroded at a 30° impact angle and V = 70 m/s along with micrographs of the three distinct scale morphologies which developed (b-d). The direction of particle impact in all figures in this section was down from the top of the photos. The center area had a comparatively smooth scale that is shown in Fig. 5.21(b); it is designated the center of the primary zone. The larger white area surrounding it had the V- shaped microstructure of the scale shown in Fig. 5.21(c); it is designated the primary zone. The white color is due to a loose layer of alumina erodent particle fines that deposited on the surface of the specimen. The third distinct scale morphology is shown in Fig. 5.21(d). It consisted of nodules of iron oxide randomly distributed over a flat, continuous surface of thin scale. It occurred outside of the primary zone in the halo zone.

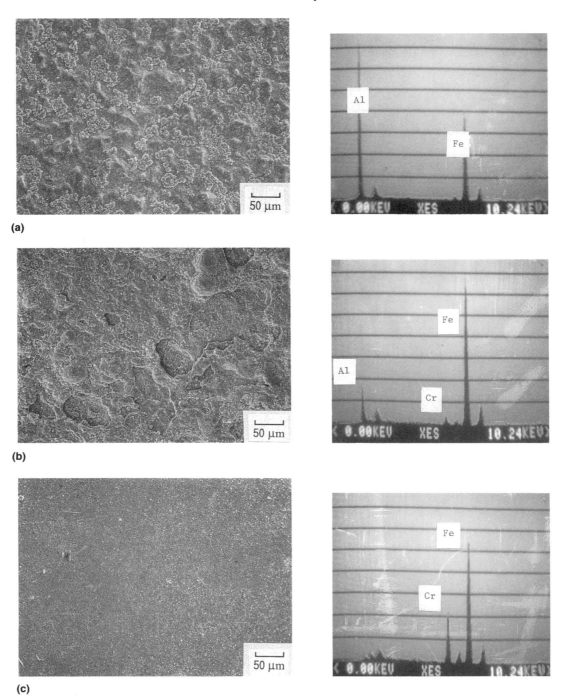

(a)

(b)

(c)

**Fig. 5.20** Scale morphology and composition in three zones of $\alpha = 90°$ test on 9Cr-1Mo steel. (a) Primary zone. (b) Halo zone. (c) Outer zone. Test conditions: nozzle tester, air, 130 μm $Al_2O_3$, $V = 35$ m/s, $T = 900$ °C, $t = 5$ h

The differences in the morphologies in the primary and halo regions of the scale on specimens eroded-corroded at 90° and 30° impact angles can be seen by comparing Fig. 5.20 and 5.21, respectively. The lower test velocity of the particles, 35 m/s, used in the 90° impact angle test is not considered to be responsible for the differences shown because surfaces of specimens tested at $V = 70$ m/s at 90° had very similar morphologies.

The macrograph of the erosion area (Fig. 5.21a) had a white region of comminuted, submicron size $Al_2O_3$ erodent covering part of the primary zone. In the center area of the primary zone where the scale surface is smoother, the $Al_2O_3$ layer spalled off on cooling. In the remainder of the primary zone, where the V-shaped scale morphology occurred, the $Al_2O_3$ layer remained. The nature and location of the $Al_2O_3$ and the ease

with which it was cleaned off the surface in an ultrasonic water bath after the test indicates that it was a loose deposit of fines that had a minimum effect on the erosion of the scale.

The center of the primary zone (Fig. 5.21b) consisted of a fairly continuous appearing scale whose surface was relatively smooth and had no observable cracks. This scale is similar to the continuous scale that occurred in the primary zone on the 90° impact test specimen, but it has more discontinuities and is considerably thinner after 5 h of testing at $V = 70$ m/s, as will be shown. This zone had the greatest concentration of particles striking the surface per unit of time.

The remainder of the primary erosion area had a distinct V-shaped notch morphology as seen in Fig. 5.21(c). This morphology was caused by the

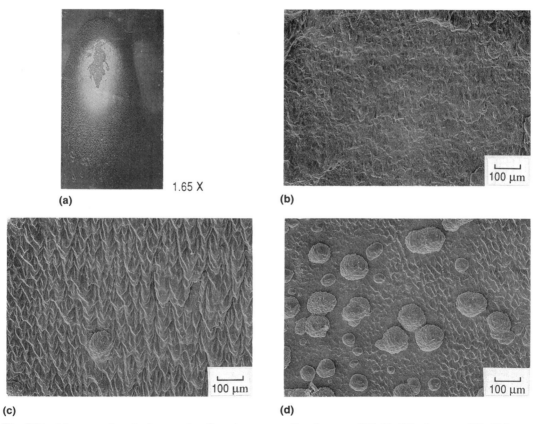

1.65 X

(a)

(b)

(c)

(d)

**Fig. 5.21** Macrograph and micrographs of erosion zone surface from $\alpha = 30°$, $V = 70$ m/s test on 9Cr-1Mo steel. (a) Eroded-corroded area. (b) Center of primary zone. (c) Primary zone. (d) Halo zone. Test conditions: nozzle tester, air, 130 μm $Al_2O_3$, $T = 850$ °C, $t = 5$ h

shallow angle of the erodent impacts breaking off V-shaped pieces of scale much as a bullet breaks out a conical-shaped area when it strikes a glass plate. In this region of the erosion area, the concentration of particles striking the surface per unit time and their velocity and impact angle have been reduced because of the fluid mechanics of the flow out of the 4.5 mm nozzle across the 1.25 cm distance separating the nozzle from the specimen surface (Ref 21).

The halo zone has still another morphology because of additional reductions in the particle concentration, velocity, and impact angle. Figure 5.21(d) of the area near the division between the primary zone and the halo zone shows some V-shaped scale areas interspersed with distinct nodules of $Fe_2O_3$ + Cr (determined by EDX peak analysis).

There is a crater resulting from a nodule that has been knocked off the surface in Fig. 5.21(c). It is gradually filling in with newly formed scale, a portion of which has some small V-shaped notches in it (at top of the crater). This occurrence and other evidence (Ref 22) indicates that nodules formed initially over the entire erosion area surface as the result of the impact of the initial particles, which damaged discrete locations of the pre-existing, thin-scale layer. Subsequently, these nodules either grew together or were knocked off by particle impacts and filled in, resulting in a more continuous scale forming in the primary zone.

In Fig. 5.22 are higher magnification micrographs of the principal scale morphologies which occurred in the 30° impact erosion tests at $V = 70$ m/s. Figure 5.22(a) shows the relatively

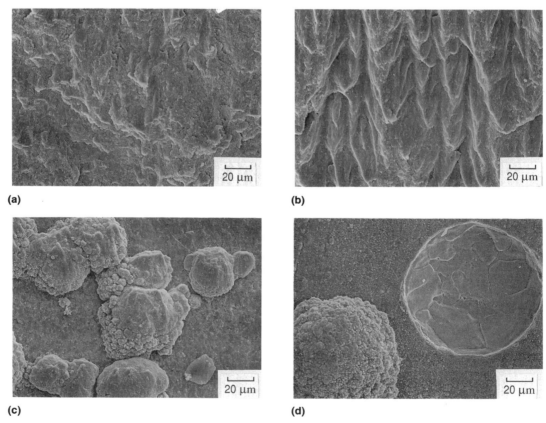

(a)

(b)

(c)

(d)

**Fig. 5.22** Micrographs of eroded scale surface from $\alpha = 30°$, $V = 70$ m/s test on 9Cr-1Mo steel. (a) Center of primary zone. (b) Primary zone near halo zone. (c) Halo zone. (d) Halo zone near outer zone. Test conditions: nozzle tester, air, 130 μm $Al_2O_3$, $T = 850$ °C, $t = 5$ h

smooth, continuous scale in the center of the primary zone with some discontinuities but no cracks. Figure 5.22(b) shows the coarse V-shaped notches that occurred in the primary zone. The difference in the topography between Fig. 5.22(a) and (b) accounts for the behavior of the alumina erodent layer that deposited on the surface. The alumina stuck better to the rougher surface during cooling and spalled off the smoother, central region of the primary zone.

Figure 5.22(c) shows nodules on a smooth base scale. Some of the nodules had small bumps on portions of them, while others appeared to have eroded areas. Some nodules had bumps over their entire surface. Figure 5.22(d) shows a nodule and an adjacent crater from which a nodule has been knocked off. The bottom of the crater had a grain pattern that reflects the grain size and shape of the underlying metal, as was observed in tests performed at an impact angle of 90°. The base scale out of which the nodules have grown was very thin, as evidenced by the still visible straight lines which are polishing marks from the metal surface preparation. Figure 5.22(d) was taken far out in the halo zone where relatively few particles had struck the surface while Fig. 5.22(c), which had a thicker scale, was taken near the primary zone where considerably more particles had struck. It has been observed earlier in this chapter that particle impacts enhance scale growth (Ref 15).

**Nodule Formation.** Figure 5.23 shows the morphology of a typical nodule and identifies the composition of the various features in its immediate area. The nodules appear to grow in a layered manner with small bumps occurring on both inside and outside layer surfaces. In some instances, parts of the bumps are apparently broken off, as can be seen in location 3, or eroded off (see Fig. 5.22c). In the case of 90° impact angle tests,

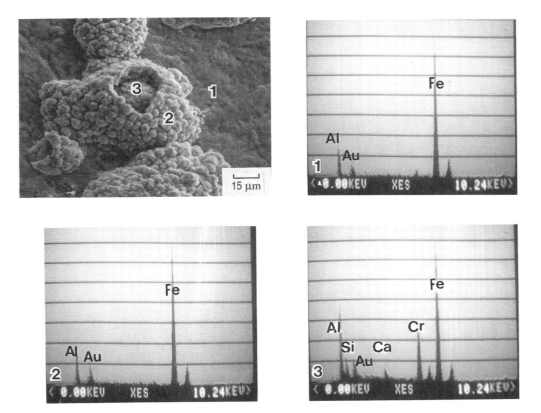

**Fig. 5.23** Nodule morphology and composition from $\alpha = 30$, $V = 35$ m/s test on 9Cr-1Mo steel. Test conditions: nozzle tester, air, 130 μm Al$_2$O$_3$, $T = 850$ °C, $t = 5$ h

the small bumps on the main mass of a nodule are eroded away, but those around the base are not.

The relatively smooth surface scale alongside the nodules that was well below the plane of the top of the nodule is $Fe_2O_3$, as indicated in EDS peak No. 1 and by x-ray diffraction. The composition of the outer surface of the nodule is $Fe_2O_3$ as indicated by peak analysis No. 2 and x-ray diffraction. A small, residual amount of submicron-size alumina erodent remains in the surface crevices of the nodule resulting in an aluminum peak. Peak analysis No. 3 of the top of the lower element of the nodule indicates that it is an iron-chromium oxide. The opening in the top of the

nodule was made prior to the end of the test exposure as some alumina is indicated to be on the surface of the inner layer.

The cross section of a fully developed nodule, unetched, is shown in Fig. 5.24, along with EDS peak analyses identifying its compositions. It can be seen that each nodule has an upper element above the general plane of the scale and a lower element that extends down into the base metal. This scale morphology has been observed by others (Ref 23-26). It appears that the nodule system must be in direct contact with the base metal to form and grow. It is not observed to grow on top of an already established, thick scale layer.

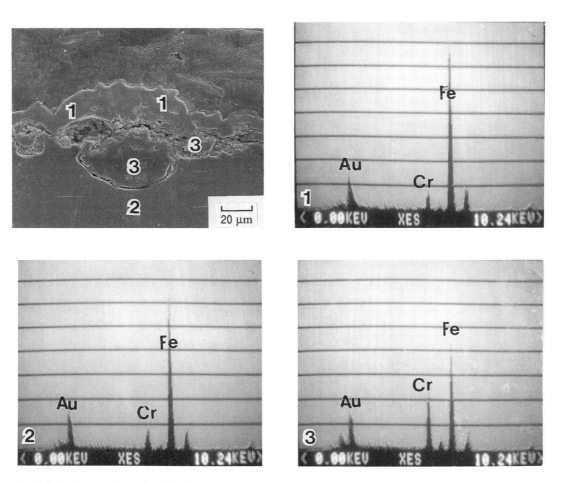

**Fig. 5.24** Cross section of nodule from $\alpha = 30°$, $V = 35$ m/s test on 9Cr-1Mo steel. Test conditions: nozzle tester, air, 130 µm $Al_2O_3$, $T = 850$ °C, $t = 5$ h

A porous, cracked scale area occurs between the two basic elements of the nodule. Some of the void areas extending into the upper element of the nodule are quite large, thinning the upper element wall enough so that pieces of the nodule surface can be broken off by impacting particles, as is seen to have occurred in Fig. 5.23. The one on the left side of the main nodule in Fig. 5.24 is an example. In most cases, the scale constituting the upper element of the nodule is relatively dense but does contain some porosity and cracks. The cross section of the bumps on the upper element

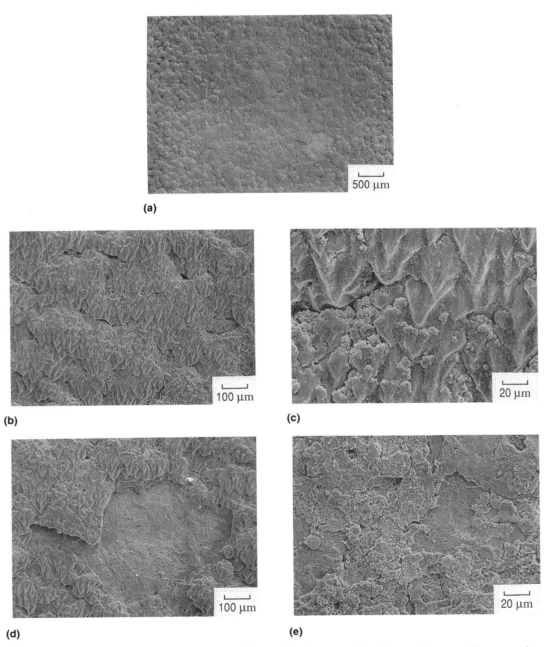

**Fig. 5.25**  Morphology of primary zone in $\alpha = 30°$, $V = 25$ m/s test on 9Cr-1Mo steel. Test conditions: nozzle tester, air, 130 µm Al$_2$O$_3$, $T = 850$ °C, $t = 5$ h

of the nodule can be readily seen. The composition of the upper element of the nodule system is $Fe_2O_3$ with none or only a small amount of chromium in it. In the nodule cross section there is no indication of $Al_2O_3$ erodent. This is an important observation, relating to when and how alumina gets and remains on the eroding surface.

The lower part of the nodule consists of a dense structure that has cracked free from the base metal. Removal of both the upper and lower elements of the nodule by particle impacts after it has cracked free from the base metal results in the smooth-walled crater shown in Fig. 5.22(d). The composition of the lower element of the nodule is iron oxide with dissolved chromium in it. The composition of the segments of scale in the crack between the base metal and the lower element of the nodule is a brittle $FeCr_2O_4$ spinel, which accounts for the crack that has formed. The

higher chromium content in the crack zone compared to the chromium content of the bulk of the lower element of the nodule shown in x-ray maps and x-ray diffraction indicated that the scale in the crack zone was $FeCr_2O_4$ and not iron oxide with dissolved chromium in it as occurred in the body of the nodule's lower element.

**Segmented Scale Formation.** The effect of a lower particle velocity on the scale morphologies of specimens tested at an impact angle of 30° can be seen in Fig. 5.25. The center of the primary zone at the lower velocity of $V = 25$ m/s in this test Fig. 5.25(a) consisted of clumps of V-shaped notch scale areas. There is no evidence of condensed, smoother surface scale in the center of the primary zone as occurred in the $V = 70$ m/s tests. The same V-shaped notches occur in $V = 70$ m/s tests but outside of the center of the primary zone.

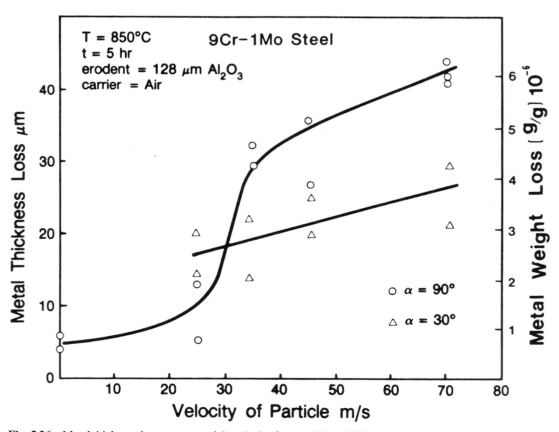

**Fig. 5.26** Metal thickness loss versus particle velocity for $\alpha = 30°$ and 90° tests

In the $V = 25$ m/s test, the scale outside of the center but still in the primary zone consists of nodules that are growing together to become a more continuous scale Fig. 5.25(b). However, it can be seen that there are discontinuities between clumps of nodules Fig. 5.25(c). These discontinuities are the reason for the formation of the segmented scales that were discussed earlier. This growth pattern of nodules of scale occurs at both 30° and 90° impact angles and is the basis for the scale loss and resultant erosion-corrosion metal loss mechanisms.

The clumps of scale in the center of the primary zone in Fig. 5.25 is not a truly continuous scale as occurs in 90° impact tests. Even at $V = 70$ m/s at 30° impact the discontinuities are present and account for the inability of the scale to build up the stresses required to cause the periodic spalling that changed the scale loss mechanism in the 90° tests (see Fig. 5.13). There is one small area in Fig. 5.25 which shows evidence of spalling Fig. 5.25(d). In this region, evidence of the underlying metal grains observed in 90° impact tests can be seen Fig. 5.25(e). Spalling was a relatively rare occurrence in the 30° impact tests at all of the velocities tested. The scale in the spalled-off region was growing as a continuous scale. The uneven surface of the scale indicates that the spall fracture plane was not at a constant level across the spalled zone.

The lack of spalling at 30° impact is another difference between the 30° and 90° impact angle tests. The morphology of the scale in the spalled area is an indication that nodules only occur once in the scale's growth history as the result of the initial impacts of individual particles on the metal surface damaging discrete locations of the pre-existing, thin scale.

**Effect of Particle Velocity.** The curves of measured metal thickness loss versus particle velocity for both the 30° and 90° impact tests are shown in Fig. 5.26. On the right hand ordinate the weight loss per gram of impacting particles was plotted. It was based on a calculation of the measured thickness loss, the eroded area and the density of the metal. The S-type transition curve for the 90° impact tests, where the scale-loss mechanism changes from a slower loss rate by

cracking and chipping to a higher loss rate by periodic spalling, does not occur at 30°. Rather, a continuous, positive slope, straight-line curve best fits the 30° impact data points. Comparing the shape of the curve to the appearance of the scale surfaces and cross sections discussed above indicates that at 30° impact angle the transition from a cracking and chipping mode of scale erosion to a spalling mechanism has not yet occurred at particle velocities up to $V = 70$ m/s. The transition occurred at a particle velocity of $V = 30$ m/s in the 90° impact tests.

The difference in the erosion mechanisms between the 30° and 90° impact tests is also indicated in Fig. 5.27. The erodent velocities of $V = 35, 45, 70$ m/s at 90° impact are above the transition velocity, while the erodent velocities are still below the transition velocity in 30° impact tests. Therefore, the metal thickness loss is greater at 90° impact than at 30°. However, at $V = 25$ m/s, both the 90° and 30° test scales are below the transition velocity, and the metal loss rate is dependent upon cracking and chipping at both impact angles. Greater amounts of scale can be chipped out in the V-shaped notches that occur at 30° impact because the actual particle impact area is magnified by the conical shape of the chip. Therefore, the metal thickness loss at $V = 25$ m/s is greater at 30° than it is at 90° impact where smaller chips of scale are formed and knocked off the surface. This behavior is unlike that of most brittle materials where the erosion at 90° is greater than that at 30°. However, other brittle materials do not have the V-shape notch surface seen on the scale in Fig. 5.21, 5.22, and 5.25.

**Summary of 30° and 90° Impact Angle Effect on Erosion-Corrosion.** While there were some similarities in behavior at the two impact angles, the 9Cr-1Mo steel essentially behaved quite differently when all test conditions were kept constant except for the change in impact angle from 30° to 90°. The similarities were:

- A duplex scale formed at both angles, consisting of an outer layer of $Fe_2O_3$ and an inner layer of iron-chromium oxide.
- At the low velocity of $V = 25$ m/s the scales at both angles eroded by a cracking and chipping mechanism.

- The impact of small, solid particles on the surface enhanced the growth rate of the scale.

- At the higher velocities, especially at $V = 70$ m/s, the scales at the center of the impact areas at both angles were consolidated, smoothed out, and had relatively few interruptions in their continuity.

The reason why there is no mechanism change at 30° impact angle while one occurred at 90° relates to the nature of the scale that formed. The vertical component of the impact force of the particles at 30° was not sufficient to overcome the scale domains that formed when the nodules grew together (Fig. 5.25) and condense the scale sufficiently to set up the stresses that could cause spalling. The surface of the scale in the center of the primary zone was flattened and condensed in

the $V = 70$ m/s, 30° impact tests, the same as occurred in the 90° tests, but apparently not to the same degree. Also, the scale thickness in the center zone of the specimens tested at $V = 70$ m/s, 30° impact was not sufficient to develop spalling stress levels. This may be due to the increased erosion rates by the cracking and chipping mechanism at 30° (Fig. 5.27) which kept the outer scale layer thinner than it was in the 90° tests.

### Correlation of Laboratory Nozzle Tester and Fluidized-Bed Combustor Erosion-Corrosion[Ref 27]

Metallographic analysis of exposed tubes from an experimental atmospheric fluidized bed combustor (AFBC) operated at Point Tupper, Nova Scotia, indicated that there were similari-

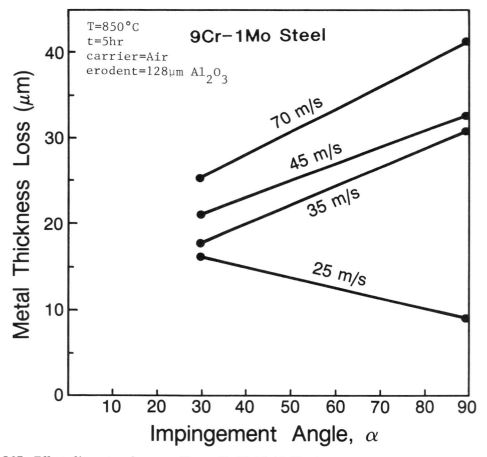

**Fig. 5.27** Effect of impact angle on metal loss at $V = 25, 35, 45, 70$ m/s

ties in the behavior of several steels to their behavior in the elevated-temperature nozzle tester (see Fig. 5.1). Table 5.4 lists the steels that were tested in the AFBC. The temperatures in the AFBC were lower than in the elevated-tempera-ture tester, and the exposure times were much longer. The composition of the particles and the carrier gases were different as were most of the other test-environment variables. However, all of the test variables in both exposures could be

**Table 5.4  Composition of AFBC tube steels**

| Alloy designation | Typical composition, wt% | | | |
|---|---|---|---|---|
| | Cr | Mo | Ni | C |
| Al carbon steel | ... | ... | ... | 0.3 |
| T22 | 2.25 | 1 | ... | ... |
| T91 | 9 | 1 | ... | ... |
| 304H | 19 | ... | 10 | ... |
| 310 | 25 | ... | 20 | ... |

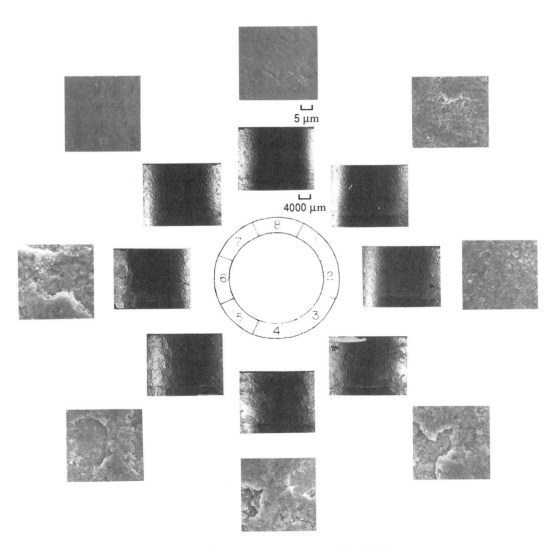

**Fig. 5.28**  Macrographs and micrographs of scale formed around T91 AFBC tube

considered as being in the same regime of low velocity, oxidizing gas particle flows.

Figure 5.28 shows macrographs and micrographs at eight locations around an AFBC tube of 9Cr-1Mo steel exposed for 5000 h at 450

to 550 °C. The distinct pattern that occurred was representative of most of the tubes in the AFBC of several different alloys. The curves of metal loss at eight positions around the periphery of several tube specimens are shown in Fig. 5.29. The pattern of metal loss is highest at the bottom,

Probe 1A1   5000 h

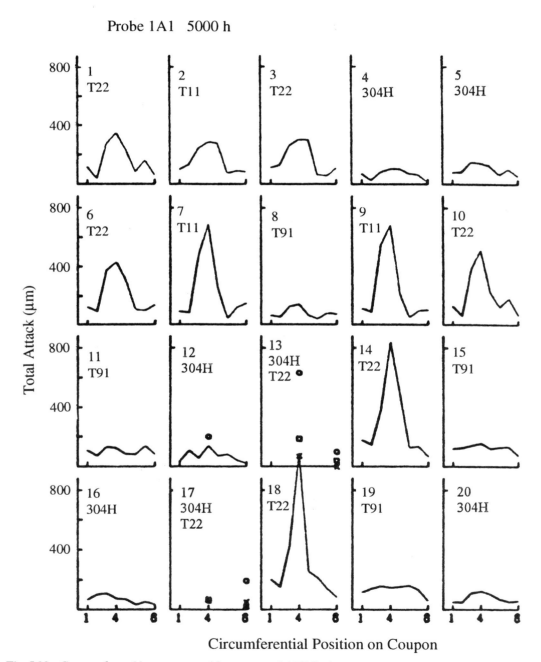

**Fig. 5.29**   Curves of metal loss versus position on several AFBC tubes

No. 4 position, and lowest at the top, No. 8 position. The scale surface in the top region (position 8) of the tube was smooth while that in the bottom region (position 4) was spalled. Metal loss rates at the bottom were generally considerably higher than those at the top at the same longitudinal tube location. At the tops of the tubes, the velocity and impact angle of the striking bed material particles were lower than those of the particles striking the tube bottoms. The surface temperatures at the tube tops were generally 50 to 80 °C higher than those on the bottoms. The lower velocity, shallow impact angle of the particles striking the top surface of the AFBC tubes related to the lower velocity and impact angle particles striking the 9Cr-1Mo steel test specimens in the erosion-corrosion laboratory tests with the same results. Scale loss was by a slow cracking and chipping mechanism that left the surface relatively smooth.

On the bottom of the tubes, the bed material particles struck at a steeper impact angle and a higher velocity, causing the scale to consolidate and removal to occur by the faster rate, spalling mechanism. This behavior related to the 90° impact angle, higher-velocity tests in the erosion-corrosion laboratory tests. Detailed microscopic observation of scales on several different tube alloys at different operating temperatures and times substantiated the similarity between the behavior of the scales on the tubes and on the specimens in the laboratory tests.

It has been consistently observed in both in-service and laboratory erosion-corrosion exposures that the scale layer that forms grows to a relatively constant thickness, morphology, and composition in a short time and remains essentially the same for thousands of hours of exposure. As scale is removed from the gas-particle impact surface by erosion, it is replaced by oxide growing from the base metal and erodent particle deposition. A plot of the short time (15 to 30 min) taken in a laboratory erosion-corrosion test for the scale layer to

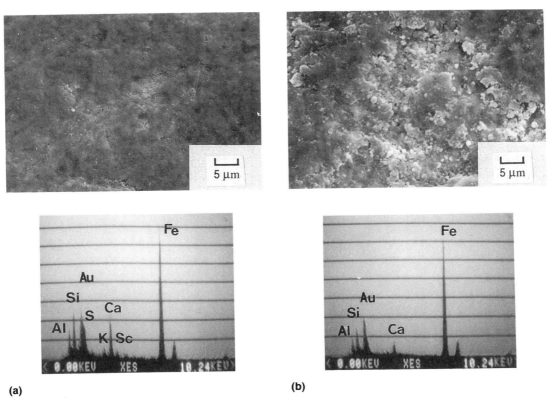

(a)                                                      (b)

**Fig. 5.30**   Surface of (a) top and (b) bottom of carbon steel tube. $T = 450$ to $550$ °C, $t = 1000$ h

reach the steady state condition is given in Ref 13.

Figure 5.30 shows typical scale morphologies at the top and bottom of a carbon steel tube section exposed for 1000 h at a test temperature of 450 to 550 °C. The scale at the top has the segmented morphology that results in scale loss by a slow cracking and chipping mechanism. The scale at the bottom at the same longitudinal location shows that it was being removed by a faster spalling mechanism. The cross section of the top and bottom scale of the carbon steel in Fig. 5.31 shows the different morphologies more clearly. The top scale (Fig. 31a and b) is segmented while the bottom scale (Fig. 31c) is more consolidated. The higher material loss rate on the tube bottom is reflected by its scale being much thinner than the scale on the top of the tube.

Similar behavior was observed for four other steels listed in Table 5.4. In every instance where the bottom of the tube had more metal loss than the top, the scale-loss mechanism on the top was cracking and chipping while that on the bottom was periodic spalling. In those instances where the top and bottom of a tube at a specific location had similar metal losses, the morphology of the top and bottom scale surfaces were essentially the same. More than 25 tube sections from the Point Tupper AFBC were metallographically examined, and the results reported in Ref 27.

Thus, while several of the conditions differed between the two types of test exposures, the basic surface degradation modes were essentially the same. Other factors such as scale composition and thickness and bed-material deposition also played a role in the behavior of the scales, but the two different scale morphologies were the major factor in the metal loss differences that occurred.

### Effect of Chromium Content of the Steels[Ref 28]

**Alumina Erodent.** Table 5.2 lists the compositions of six alloys that were tested. The effect of the chromium content of steels on their ero-

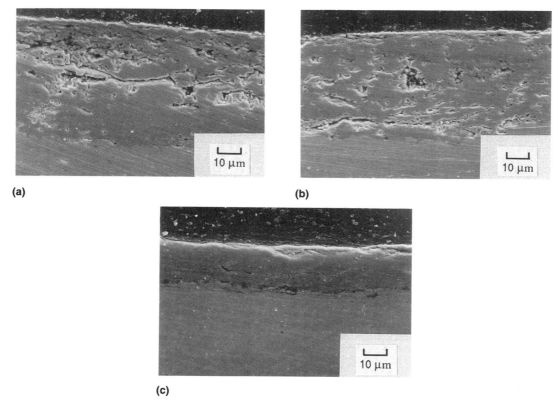

**(a)**

**(b)**

**(c)**

**Fig. 5.31** Cross section of (a), (b) top and (c) bottom of carbon steel tube. $T = 450$ to $550$ °C, $t = 2000$ h

sion-corrosion behavior has the same basic pattern that occurs in static oxidation. However, the morphology and composition of the scales formed and the mechanisms and rates of the weight changes that occurred were significantly affected by the impact of particles on the growing scale. Figures 5.32 and 5.33 show the surfaces of the six chromium steels after 5 h of dynamic

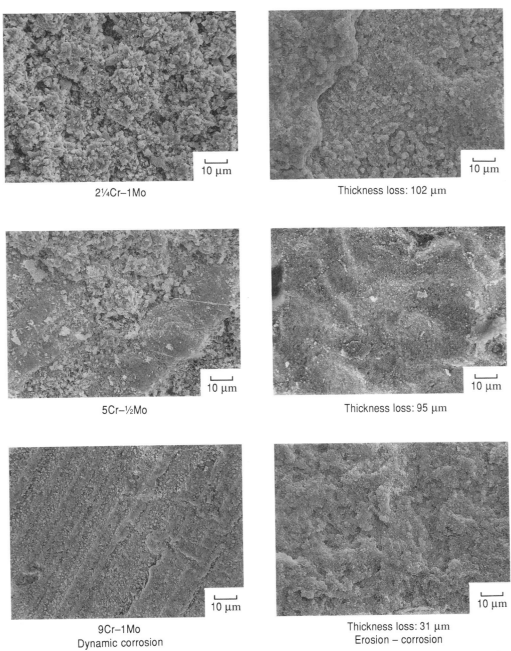

2¼Cr–1Mo

Thickness loss: 102 μm

5Cr–½Mo

Thickness loss: 95 μm

9Cr–1Mo
Dynamic corrosion

Thickness loss: 31 μm
Erosion – corrosion

**Fig. 5.32** Surface of 2¼Cr, 5Cr, and 9Cr steel after dynamic corrosion at $V = 50$ m/s and combined erosion-corrosion at $V = 35$ m/s with 130 μm $Al_2O_3$. Test conditions: nozzle tester, air, $\alpha = 90°$, $T = 850$ °C, $t = 5$ h

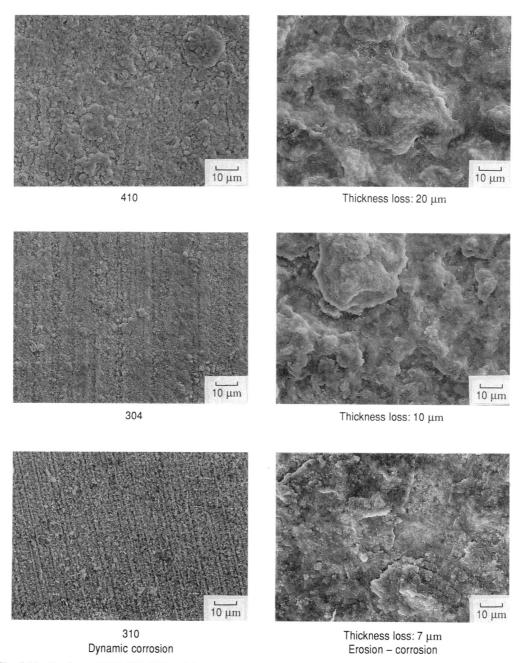

**Fig. 5.33** Surface of 410, 304, 310 stainless steels after dynamic corrosion at $V = 50$ m/s and combined erosion corrosion at $V = 35$ m/s with 130 μm Al$_2$O$_3$. Test conditions: nozzle tester, air, $\alpha = 90°$, $T = 850$ °C, $t = 5$ h

corrosion (left-side micrographs) and erosion-corrosion (right-side micrographs). As the chromium content of the steel increased, the

morphology of the scale changed significantly. The 2¼ and 5Cr steels that were dynamically corroded (no particles in the gas flow) developed

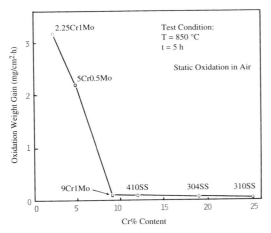

**Fig. 5.34** Static furnace corrosion of chromium steels at 850 °C

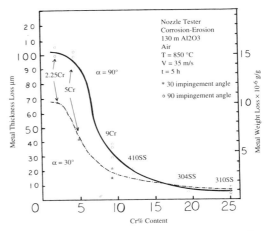

**Fig. 5.35** Erosion-corrosion metal loss curves of chromium steels at 850 °C

thick, segmented scales. They developed much smoother, consolidated scales when impacting particles and corrosive gas affected the surface simultaneously.

For the scales with more than 9% Cr, dynamic oxidation resulted in a thin scale layer of $Cr_2O_3$, which still showed the polishing lines on the base metal. When these steels were impacted with particles during oxidation (right-hand micrographs), the morphology and thickness as well as the composition of the scales changed. X-ray diffraction of the scales indicated that the 9Cr-1Mo steel formed an iron-oxide scale while the three higher chromium content steels had $FeCr_2O_4 + Cr_2O_3$ scales. The scales became more segmented as the chromium content increased.

The weight change curve in Fig. 5.34 shows the weight gain that occurred as the result of scale growth after static furnace oxidation for a 5 h, comparable time to the erosion-corrosion exposure time. Even though the curve represents the results of static oxidation, the morphology of the dynamic oxidation scales in Fig. 5.32 and 5.33 can be related to it. The 2¼ and 5Cr steels, which formed thick, iron oxide scales, show a high weight gain, while the 9Cr and higher chromium steels, which all formed a thin, protective $Cr_2O_3$ scale, show nearly the same, very low weight gain.

Combined erosion-corrosion resulted in the curves shown in Fig. 5.35, which have the same general shape as the static corrosion curve in Fig. 5.34. However, there are differences in the erosion-corrosion curves from the static corrosion curve that are the result of the impact of erodent particles on the growing scale. These changes can be related to the scale morphologies shown in Fig. 5.32 and 5.33. The static corrosion curve shows a weight *gain* as the undisturbed scale grew on the metal substrate, while the erosion-corrosion curves show a weight *loss*. The static air exposure resulted in a thin $Cr_2O_3$ scale forming on the 9Cr-1Mo steel in the short, 5 h exposure that was similar to the scales that formed on the higher chromium steels. The resulting weight gain for the 9Cr steel in Fig. 5.34 is the same low rate as for the higher chromium content steels.

When particles struck the surface of the 9Cr steel, they caused a relatively thick $Fe_2O_3$ scale to form and the metal loss rate of the 9Cr steel increased above that of the higher chromium steels (Fig. 5.35). Another difference that was caused by the erodent particles was the two distinct metal loss curves, which could be plotted as a function of particle impact angle. Thus, while erosion-corrosion curves are similar in shape to static corrosion curves, there are distinct differences that indicate that the impacting particles modify the oxidation behavior of the chromium steel (see Fig. 5.11).

A more detailed analysis of the erosion-corrosion curves of Fig. 5.35 indicate several aspects of the behavior of the scale. At an impact angle of 90°, the consolidated scale of the 2¼ and 5Cr steels was removed by the same periodic spalling mechanism and there was little difference between their metal loss rates. This differs from their static corrosion behavior where the 2¼ Cr steel had a greater weight change than the 5Cr steel. The tests carried out at 30° impact angle showed a much larger difference in metal loss between the two steels and an overall much lower loss than occurred at 90°.

The more segmented and thinner scales on the higher chromium steels resulted in much lower material losses because of the change in the scale-loss mechanism and the thinner scales that formed. For the $Cr_2O_3$ + spinel scales that formed on stainless steel types 410, 304, and 310, the metal losses were comparable, both as a function of chromium content and impact angle. These thin scales transmitted much of the force of the impacting particles through to the base metal, making the conditions to remove scale approximately the same.

The morphology of the cross sections of two types of scale that formed and were removed in the erosion-corrosion tests are shown in Fig. 5.36 and 5.37. The thick, consolidated scale on the 2¼Cr-1Mo steel is shown in Fig. 5.36. Its surface morphology is seen in Fig. 5.36(a), an area where the scale had completely spalled off in Fig. 5.36(b), and an area of dense scale layers in Fig. 5.36(c). In contrast, the 310 scale in Fig. 5.37 shows the thin, continuous $Cr_2O_3$ that formed on the furnace-oxidized specimen and the changing thickness and state of segmentation that formed in combined erosion-corrosion tests as the test

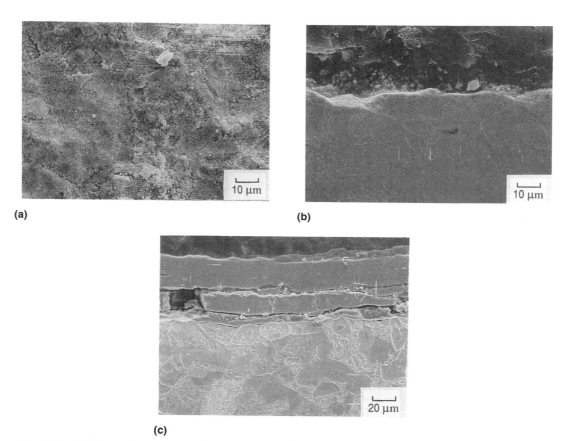

(a)

(b)

(c)

**Fig. 5.36** Surface and cross section of scale formed on 2¼Cr-1Mo steel. Test conditions: nozzle tester, air, 130 μm $Al_2O_3$, α = 30°, T = 850 °C, V = 35 m/s, t = 5 h

conditions became more severe. The 70 m/s test scale in Fig. 37(d) clearly shows the segmented morphology of a thicker scale.

**Silicon Carbide Erodent.** The alumina erodent used in the previously described tests was agglomerated and rounded in configuration. When angular silicon carbide erodent was used, the erosion-corrosion behavior of the steels was different. Figure 5.38 relates the behavior of the steels eroded by silicon carbide to those eroded by $Al_2O_3$. It plots the weight loss for the 9Cr-1Mo steel using angular SiC and round $Al_2O_3$ erodents. The lower curves are similar to the curves plotted in Fig. 5.26, the only difference being the test time, 2 versus 5 h. The great increase in erosivity of the silicon carbide particles is primarily due to their angular shape, compared to the rounded shape of the $Al_2O_3$ particles, as

discussed in Chapter 4 (Ref 29). The sharp particles removed scale as it formed, resulting in metal rather than scale being eroded at the 850 °C test temperature. At this high a temperature, the metal erosion rate was much higher than the rate of loss of the scale in the $Al_2O_3$ erodent tests (Ref 30).

The effect of chromium content on weight loss of a number of alloys impacted with silicon carbide particles is shown in Fig. 5.39. The overall shape of the curves of weight loss versus chromium content at 850 °C was similar to the alumina erodent curves at 850 °C (see Fig. 5.35) but a somewhat different shape occurred in tests carried out at 650 °C. At the lower test temperature there was relatively little difference in weight loss as a function of chromium content. For the higher-chromium steels, there was little difference in weight loss at both test temperatures.

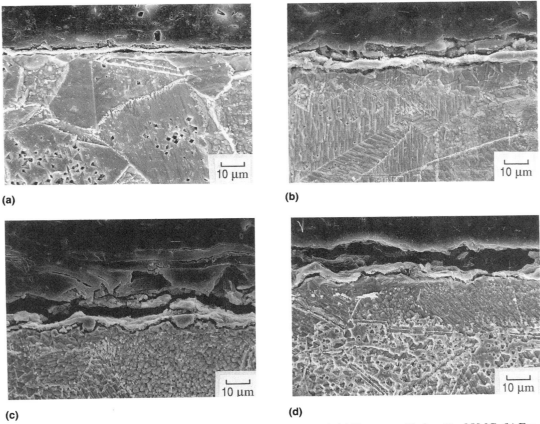

(a)

(b)

(c)

(d)

**Fig. 5.37**  Cross section of scale formed on type 310 stainless steel. (a) Furnace oxidation, $T = 850$ °C. (b) Erosion-corrosion, $V = 35$ m/s, $\alpha = 30°$, $T = 850$ °C. (c) Erosion-corrosion, $V = 35$ m/s, $\alpha = 90°$, $T = 850$ °C. (d) Erosion-corrosion, $V = 70$ m/s, $\alpha = 90°$, $T = 850$ °C

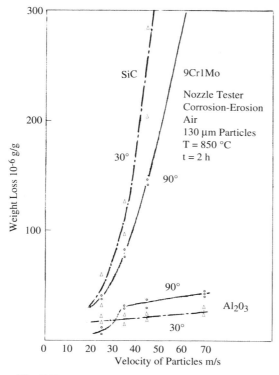

Fig. 5.38 Weight loss versus particle velocity for Al₂O₃ and silicon carbide erodents

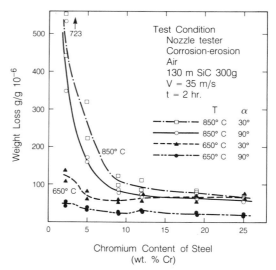

Fig. 5.39 Weight loss of chromium steels eroded-corroded using silicon carbide erodent

The differences and similarities in weight loss as a function of chromium content and test temperature can be related to the surface morphology of the alloys. The steels shown in Fig. 5.40 that were tested at 650 °C (left-hand column) generally had no scale layer during erosion at the 30° impact angle, except for the 2¼Cr-1Mo steel, which had a thin, intermittent layer of scale. The absence of scale is evidenced by the metal platelets that formed and the deep narrow gouges that occurred (see also Fig. 5.7a). The 30° impact angle weight loss curve at 650 °C in Fig. 5.39 indicates a somewhat higher weight loss for the 2¼Cr steel compared to the others whose rates are nearly the same.

At 850 °C, the 2¼Cr-1Mo steel has an established, consolidated scale layer, see Fig. 5.40, that lost material at a high rate by periodic spalling, as can be seen in the 30° impact angle curve in Fig. 5.39. The higher chromium steels had decreasing amounts of intermittent scale with resul-

tant weight losses that were much lower and more comparable between alloys.

## Reducing Atmosphere Erosion-Corrosion[Ref 31, 32]

A large, long-duration investigation of the elevated-temperature corrosion and erosion-corrosion of materials operating in reducing gas coal gasification atmospheres (CGA) was carried out at IITRI that produced considerable knowledge that is applicable to materials degradation in a number of carbon base fuel combustion systems where reducing or alternating oxidizing and reducing atmospheres occur. The program, carried out over a 10-year period, utilized large, sealed-chamber complex, multispecimen combined erosion-corrosion test equipment (Ref 6, 7, 31) that operated over a wide range of elevated-temperature conditions at both ambient and high pressures. A variety of sulfur-containing reactive gases were used along with a wide range of erodent-particle materials and dynamic conditions, and exposure times from 50 to 250 h. The erosion-corrosion tests were augmented with up to 10,000 h corrosion-only tests in the same gaseous atmospheres. The total program utilized over 10,000 exposure hours and over 100,000 specimen hours of testing. The observed behavior of alloys exposed to reducing, sulfur-containing atmospheres was, in several aspects, different from that which occurs in the generally oxidizing, combustion gases of fluid-bed combustors, pulverized-coal boilers, and other similar equipment. However, where locally reducing atmospheres occur in carbon fuel combustion zones, such as near coal-feed ports, the surface degradation can be more similar.

### Corrosion

CGA atmospheres have the potential to oxidize, sulfidize, and carburize high-temperature alloys (Ref 32). Among many variables that influence the corrosion behavior of engineering alloys are gas composition, alloy chemistry, time, temperature, pressure, and initial surface condition. Low and medium Btu CGA atmospheres are reducing atmospheres characterized by low-oxygen and high-sulfur potentials. Under these conditions, many engineering alloys may undergo severe to catastrophic corrosion during extended exposures.

Many common engineering alloys display good high-temperature corrosion resistance to oxidation because of the formation of protective-oxide scales by selective oxidation of chromium. Under multi-oxidant CGA atmospheres, Fe-Cr, Fe-Ni-Cr, and Co-Cr alloys undergo a transition from the formation of protective-oxide scales to nonprotective sulfide slags with increasing sulfur activity and temperature. Systematic studies (Ref 33) of corrosion behavior of engineering alloys in CGA atmospheres have established that material degradation is essentially determined by base-metal sulfidation which, in turn, is largely controlled by equilibrium partial pressures of oxygen ($P_{O_2}$) and ($P_{S_2}$). The presence of carburizing gases such as CO and $CH_4$ has no apparent effect under normal circumstances, but carburization may become a problem at higher pressures where carbon activities increase significantly and also if protective chromium oxides

are damaged and/or liquid corrosion products are formed.

Thermodynamically, the critical factor is stability of base metal (iron, nickel, and cobalt) sulfides or base metal-chromium sulfides in low $P_{O_2}$-high $P_{S_2}$ CGA atmospheres. Comprehensive reviews of reactions in mixed oxidants (Ref 34) stress the point that both thermodynamic and kinetic factors must be considered in the analysis of CGA corrosion behavior of materials. Thermodynamics govern the reaction paths that may be taken and the sequence of reaction products that may be formed, whereas the kinetics of the various competing processes establish a rate-determining step that dictates the actual products that will form and the useful life that can be realized. Therefore, thermodynamic phase stability diagrams have only limited predictive capability in multi-oxidant, complex, CGA environments.

### Erosion-Corrosion

In the case of kinetically dominated, complex dynamic processes such as CGA erosion-corrosion, the utility of equilibrium thermodynamic phase stability diagrams is limited even more so to providing only qualitative guidelines. The chemistry of high-temperature CGA erosion-corrosion is extremely complex. It involves not only the reactions at the gas/metal and scale/metal interfaces but also continuous interaction and competition between erosion and corrosion kinetic processes. In addition, reactive erodents may give rise to highly reactive erosion-corrosion debris (for example, by slagging) which, in turn, may flux protective surface oxide scales and attack the sound metal underlying these reaction products.

### Test Materials and Conditions

The parameters studied in the program were particle velocity, erodent concentration (solids loading), temperature, gas pressure, erodent type, particle size, erodent hardness, sulfur compounds (in gas and erodent), impact angle, exposure time, and alloy composition.

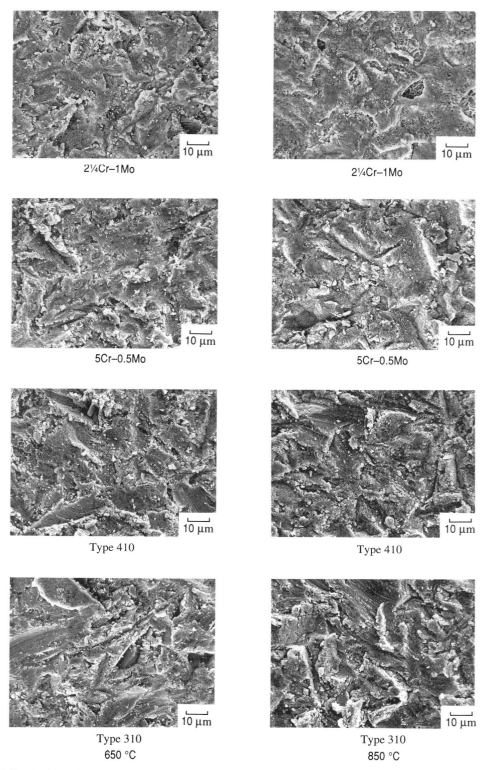

2¼Cr–1Mo

2¼Cr–1Mo

5Cr–0.5Mo

5Cr–0.5Mo

Type 410

Type 410

Type 310
650 °C

Type 310
850 °C

**Fig. 5.40** Surface of chromium steels eroded-corroded using silicon carbide erodent. Test conditions: nozzle tester, air, 130 μm SiC, α = 30°, V = 35 m/s, t = 2 h

Most of the tests were run at 15 or 35 m/s velocity and lasted either 50 or 100 h. Except for a few selected runs, particle loading remained at 0.003 lb/ft$^3$ and the impact angle was kept constant at 45°. Although some intermediate temperatures and pressures were included in the test matrix, the majority of runs were conducted at 900 or 978 °C and either atmospheric or 1000 psi pressure. A reducing-gas atmosphere occurred in all tests in this investigation as the sulfur level was set to simulate that in coal-gasification systems.

The alloys tested are listed, along with their compositions, in Table 5.5. The majority of tests employed real coal char erodents obtained from coal gasification pilot plants. About one-third of the tests used metallurgical coke for the erodent. Unlike essentially inert erodents such as alumina or dolomite (used in a few tests), both the chars and coke contained varying amounts of chemically reactive substances, principally sulfur compounds. The constituents causing erosion were various ash minerals, predominantly silica, aluminosilicates and pyrites.

Most of the tests were conducted with fairly coarse erodents (−20+50 mesh); however, several runs utilized fine fractions (−200+400 mesh) to study the effect of particle size. Erodents were

**Table 5.5   Materials in long-duration tests**

| Alloy | Symbol | Fe | Ni | Cr | Mo | Co | W | Others |
|---|---|---|---|---|---|---|---|---|
| Type 310 | 310 | bal | 20.5 | 25 | ... | ... | ... | ... |
| Type 329 | 329 | bal | 4.5 | 27.5 | 1.5 | ... | ... | ... |
| Type 446 | 446 | bal | ... | 25 | ... | ... | ... | ... |
| HK-40 | HK-40 | bal | 20 | 26 | ... | ... | ... | ... |
| HL-40 | HL-40 | bal | 20 | 30 | ... | ... | ... | ... |
| Incoloy 800(a) | 800 | bal | 32.5 | 21 | ... | ... | ... | 0.4 Al, 0.4 Ti |
| Sanicro 32X(b) | 32X | bal | 31 | 21 | ... | ... | 3 | 0.4 Al, 0.4 Ti |
| Crutemp 25(c) | CRU-25 | bal | 25 | 25 | ... | ... | ... | ... |
| LM-1866(d) | LM-1866 | bal | ... | 18 | ... | ... | ... | 6 Al, 0.6 Hf |
| Inconel 601(a) | 601 | bal | 60.5 | 23 | ... | ... | ... | 1.4 Al |
| Inconel 671(a) | 671 | ... | bal | 48 | ... | ... | ... | 0.4 Ti |
| Hastelloy X(e) | H-X | 18.5 | bal | 21.7 | 9 | 1.5 | 0.6 | ... |
| RA-333(f) | RA-333 | bal | 45 | 25 | 3 | 3 | 3 | ... |
| Supertherm(g) | T63WC | bal | 36 | 27 | ... | 15 | 5 | ... |
| Wiscalloy 30/50W(h) | 30/50W | bal | 47.5 | 28 | ... | ... | 3.6 | ... |
| Multimet(e) | N155 | bal | 20 | 21 | 3 | 20 | 2.5 | 1 Nb + Ta |
| Haynes 150(e) | 150 | 18 | 1 | 27 | ... | bal | ... | ... |
| Haynes 188(e) | 188 | 3 max | 22 | 22 | ... | bal | 14 | ... |
| Stellite 1(e) | S-1 | 3 max | 3 max | 30 | 1 max | bal | 12 | 2.5 C |
| Stellite 6B(e) | 6B | 3 max | 3 max | 30 | 1.5 max | bal | 4.5 | 1.2 C |
| Aluminized 310(i) | 310AL | ... | ... | ... | ... | ... | ... | ... |
| Aluminized 800(i) | 800AL | ... | ... | ... | ... | ... | ... | ... |

(a) Inco Alloys. (b) Sandvik Steels. (c) Colt Industries. (d) Lockheed. (e) Cabot. (f) Rolled Alloys. (g) Abex. (h) Wisconsin Centrifugal. (i) Pack aluminized by IITRI

**Table 5.6   Gas composition of test atmosphere, vol%(a)**

| Component | Inlet, at ambient temperature | Equilibrium(b) 1500 °F | Equilibrium(b) 1650 °F | Equilibrium(b) 1800 °F |
|---|---|---|---|---|
| H$_2$ | 24 | 23 | 27 | 31 |
| CO$_2$ | 12 | 19 | 17 | 15 |
| CO | 18 | 11 | 14 | 17 |
| CH$_4$ | 5 | 9 | 6 | 3 |
| NH$_3$ | 1 | 1 | 1 | 1 |
| H$_2$O | bal | bal | bal | bal |
| H$_2$S | 0-1 | 0-1 | 0-1 | 0-1 |

(a) MPC standard Coal Gas Atmosphere (CGA), representative of medium Btu coal gas. (b) Calculated equilibrium composition

examined by SEM to give a qualitative characterization of particle angularity, porosity, and surface texture, all of which influence erosivity.

Except for one run in a nitrogen atmosphere, all tests employed a standard gas chemistry typifying medium Btu coal gas. Inlet and equilibrium compositions of the CGA gas mixture are shown in Table 5.6. The 0.1 to 1.0% range of $H_2S$ represents realistic compositions for medium Btu coal gas derived from domestic coals. For study purposes, a few tests were run in hydrogen sulfide ($H_2S$)-free gas mixtures. Both inlet and exit gas compositions were monitored with on-line gas chromatography. A summary of test results is contained in Table 5.7.

### Effect of Particle Velocity

The effect of particle velocity on erosion-corrosion with coke erodent (Fig. 5.41) shows a consistent and systematic increase of erosion-corrosion with velocity.

### Effect of Temperature

As illustrated in Fig. 5.42, there is a sharp increase of erosion-corrosion metal loss with temperature which, for 310 and 446 stainless steels, amounts to nearly one order of magnitude between 480 °C (900 °F) and 978 °C (1800 °F). Steeper acceleration of attack rates at higher temperature levels, typically above 875 °C (1600 °F), was observed that generally occurs in all coal-gasification environments. It is associated with excessive rates of sulfidation suffered by nearly all iron, nickel, and cobalt alloys operating in reducing or near-reducing atmospheres, resulting from the instability of protective-oxide scales. The formation and incursion of sulfides subverts the physical integrity of the scale, making it progressively more prone to erosion and spalling damage by the impinging particles. In extreme instances, damage is further intensified by formation of liquid phases (metal-sulfide eutectics), which can lead to complete destruction by alloy meltdown.

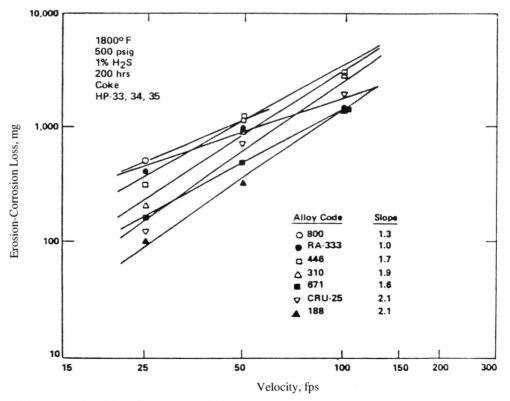

**Fig. 5.41** Effect of particle velocity on material loss in reactive gas tests

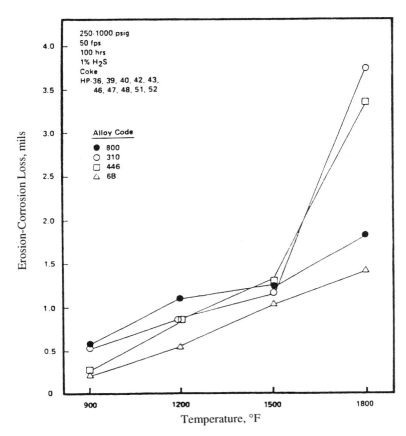

**Fig. 5.42** Effect of temperature on material loss in reactive gas tests

**Table 5.7 Summary of erosion-corrosion data for 1500 to 1800 °F tests with coal chars(a)**

| Alloy class | Alloy | Erosion-corrosion(b), mils | Cr content, wt% | Ni + Co content, wt% |
|---|---|---|---|---|
| High nickel | H-X | 250 | 21.7 | 49 |
| | 601 | 232 | 23 | 61 |
| | RA-333 | 15 | 25 | 48 |
| | 30/50W | 29 | 28 | 47.5 |
| Fe-Cr-Ni | 800 | 201 | 21 | 32.5 |
| | 32X | 154 | 21 | 31 |
| | HK-40 | 86 | 26 | 20 |
| | CRU-25 | 85 | 25 | 25 |
| | T63WC | 54 | 27 | 51 |
| | 310 | 25 | 25 | 20.5 |
| | HL-40 | 26 | 30 | 20 |
| Ni-Co-Cr | N155 | 224 | 21 | 40 |
| | 150 | 131 | 27 | 40 |
| | 188 | 76 | 22 | 62 |
| Cr-Ni | 671 | 7.6 | 48 | 51 |
| Fe-Cr | 446 | 7.8 | 25 | 0 |
| | 329 | 2.0 | 27.5 | 4.5 |
| Fe-Cr-Al | LM-1866 | 9.1 | 18 | 0 |
| Co-Cr-W | 6B | 1.3 | 30 | 60 |
| | S-1 | 1.1 | 30 | 53 |
| Aluminized Fe-Cr-Ni | 310AL | 5.4 | 25(c) | 20.5(c) |
| | 800AL | 2.9 | 21(c) | 32.5(c) |

(a) 0.3 to 1000 psig, 100 ft/s, 50 h (normalized), 1% $H_2S$, coal chars. (b) Maximum erosion-corrosion loss among all runs. (c) Base metal composition

### Effect of Pressure

The effect of gas pressure on erosion-corrosion is complex based on tests with alumina erodent summarized in Fig. 5.43. Overall, there is a direct correlation, such that all materials exhibited greater attack at 1000 psig than at 0.3 psig. The harmful effect of pressure appears to be more accentuated with materials inherently more vulnerable to attack than with alloys that demonstrated good resistance. Interestingly, the data suggest that some alloys may suffer maximum attack rates at intermediate pressures; this trend emerges even more clearly from tests employing coke erodent. No ready explanation offers itself to rationalize this complex relationship between pressure and erosion-corrosion attack.

It is difficult to establish whether pressure dependence is primarily a physical or chemical effect. Inasmuch as there is direct correspondence between gas pressure and density, it might be argued that denser gas jets may be partly responsible for intensified erosion at elevated pressures. Based on thermodynamics, higher pressures are not expected to cause increased corrosion. Kinetics, however, offer a more plausible explanation inasmuch as elevated pressures accelerate the rate of chemical reactions and diffusion phenomena.

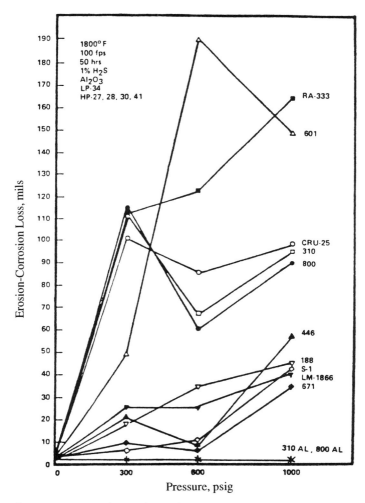

**Fig. 5.43** Effect of pressure on material loss in reactive gas tests

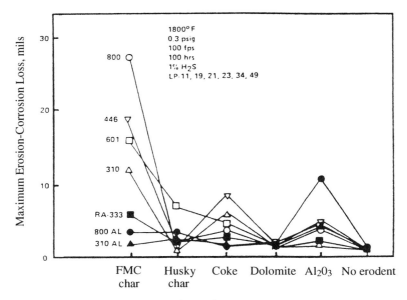

**Fig. 5.44** Effect of erodent type on material loss in reactive gas tests

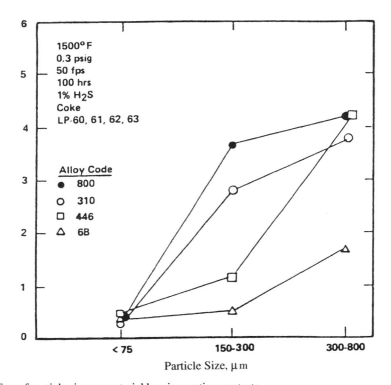

**Fig. 5.45** Effect of particle size on material loss in reactive gas tests

### Effect of Erodent Characteristics

There is ample evidence from laboratory studies as well as field experience in coal gasifiers that erosion damage is closely affected by erodent characteristics and properties, including particle size, strength, angularity, density, porosity, surface texture, and chemical reactivity (see Chapter 4). When both reactive gases and erodents are used in erosion-corrosion tests, chemical composition becomes an even more important factor.

Figure 5.44 shows how erosion-corrosion varied with some of the erodents employed in the test program. Broadly speaking, materials that generally had good surface resistance (such as aluminized 310 stainless steel) performed equally well with all erodents. The aggressive nature of FMC char relative to other erodents is attributable to

greater chemical reactivity resulting from its fairly high content of volatile sulfur species. Physical properties apparently play a secondary role, considering that some erodents were more angular (husky char), and others distinctly denser and stronger ($Al_2O_3$).

Results obtained from comparative testing with coarse- and fine-particle coke, shown in Fig. 5.45, indicate that large particles caused appreciably greater erosion-corrosion losses. This finding is consistent with results obtained from erosion studies described in Chapter 4. The large, alloy-dependent effect shown in Fig. 5.45 strongly suggests that additional factors played an influential role in the material losses plotted.

The effect of erodent strength/integrity is qualitatively illustrated in Fig. 5.46. Alumina is seen to be consistently more damaging than coke,

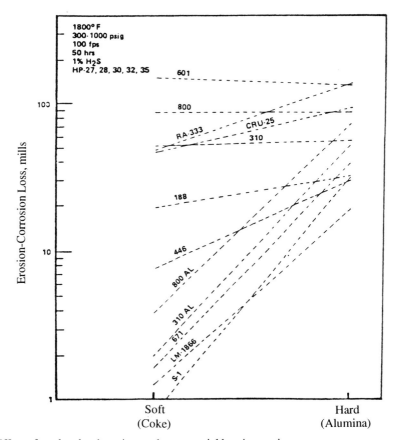

**Fig. 5.46** Effect of erodent hardness/strength on material loss in reactive gas tests

except with the most vulnerable alloys. The relative ranking of alloys and the very large spread of material loss by soft, but chemically reactive coke erodent is evidence for the strong influence of chemical effects. With alumina, the spread narrows considerably, demonstrating that the erosion contribution is significant.

### Effect of Sulfur Compounds

The presence of corrosive sulfur compounds was found to have a stimulating effect on erosion-corrosion. Sulfidation is caused not only by hydrogen sulfide in the test atmosphere but also by reactive sulfur compounds in the char and coke erodents. Both of these effects are illustrated in Fig. 5.47, which shows a general trend of increasing material loss with gas-$H_2S$ content as well as with char-sulfur content. The combination of high sulfur levels in both gas and char seems to be particularly harmful.

### Effect of Exposure Time

In the study on the effect of test duration on erosion-corrosion, one series of experiments utilized interrupted 200 h runs, with specimen weighings every 50 h. The results presented in Fig. 5.48 show constant or accelerating attack rates with exposure time. This contrasts sharply with the behavior of these alloys in simple reducing gas phase exposures where corrosion rates typically level off due to spontaneously formed protective scales.

The compound effect of corrosion and erosion apparently interferes with the regeneration of such scales, and leads to so-called "breakaway" attack that commonly occurs in reducing, sulfidizing atmospheres. It is characterized by the sudden and often drastic increase in attack exhibited by the 446 and 310 stainless steel specimens in Fig. 5.48. Because all iron-, nickel-, and cobalt-base alloys are thermodynamically unstable in

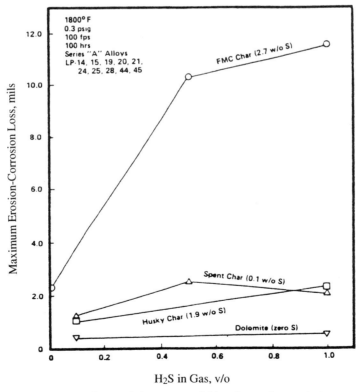

**Fig. 5.47**   Effect of sulfur content of gas and char on material loss in reactive gas tests

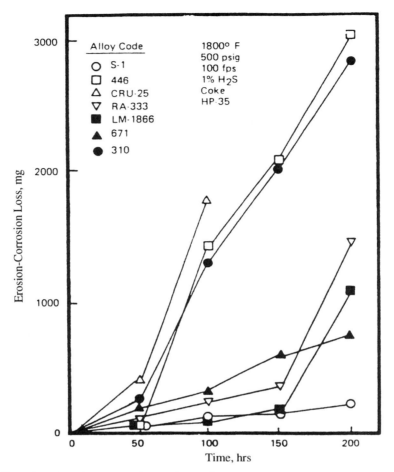

**Fig. 5.48**  Effect of test duration on material loss in reactive gas tests

the high-temperature, sulfidizing environments under discussion, breakaway effects (catastrophic sulfidation) are a normal expectation. However, there was a salient difference between corrosion and erosion-corrosion, with regard to the stability and staying power of protective scales.

In the corrosion tests, breakaway was typically encountered only after prolonged periods of exposure lasting several thousand hours, if at all. In sharp contrast, erosion-corrosion breakaway took place in as little as 50 h, shown by the curves in Fig. 5.48. As an example of the greatly accelerated onset of breakaway under erosion-corrosion conditions, Alloy LM-1866 suffered breakaway after a mere 150 h, whereas exposure in identical gas atmospheres tests produced this effect only after 5000 h.

The very early onset of breakaway in the erosion-corrosion exposures demonstrates that time is an extremely important parameter in assessing materials performance in reducing, sulfidizing atmospheres. This is not true in oxidizing atmospheres (see Chapter 6).

### References

1.  Agarwal, A., and Howes, M.A.H., Erosion/Corrosion of Materials in High-Temperature Environments, *Proc. AIME Conf. High Temperature Corrosion in Energy Systems* (Detroit), Sept 1984

2.  Nagarajan, V., and Wright, I.G., Influence of Oxide Scales on High Temperature Corrosion-Erosion Behavior of Alloys, *Proc. NACE Conf. High Temperature Corrosion*, NACE-6 (San Diego), March 1981, p 398-405

3.  Tiainen, T., et al, Simulation of the Erosion and Corrosion of Materials in Fluidized Bed Combustors, *Proc. AIME Conf. High Temperature Corrosion in Energy Systems* (Detroit), Sept 1984

4. Kliest, D.M., "One-Dimensional, Two Phase Particulate Flow," M.S. thesis, Report LBL-6967, Lawrence Berkeley Laboratory, University of California, Berkeley, 1977

5. Levy, A.V.; Aghazadeh, M.; and Hickey, G., "The Effect of Test Variables on the Platelet Mechanism of Erosion," *Wear*, Vol 108 (No. 1), Mar 1986, p 23-42

6. Levy, A.V., The Erosion of Metal Alloys and Their Scales, *Proc. NACE Conf. Corrosion-Erosion-Wear of Materials in Emerging Fossil Energy Systems* (Berkeley, CA), Jan 1982, p 298-376

7. Stringer, J., Performance Limitations of Electric Power Generating Systems Imposed by High Temperature Corrosion, *High Temp. Technol.*, Vol 3 (No. 3), Aug 1985, p 119-141

8. Natesan, K., Oxygen-Sulfur Corrosion of Metals in Mixed Gas Atmospheres, *Proc. NACE Conf. Corrosion-Erosion-Wear of Materials at Elevated Temperatures* (Berkeley, CA), Jan 1986, p 1-26

9. Minchener, A., The Performance of Materials in Fluidized Bed Combustion Environments, *Proc. NACE Conf. Corrosion-Erosion-Wear of Materials at Elevated Temperatures* (Berkeley, CA), Jan 1986, p 152-160

10. Sethi, V.K., et al, Corrosion Performance of Alloys in TVA's 20 MW AFBC Pilot Plant, *Proc. TMS-AIME Conf. High Temperature Corrosion in Energy Systems* (Detroit), Sept 1984, p 143-160

11. Nagarajan, V., Wright, I.G., and Smith, R.D., "Morphology of Corrosion of Advanced Heat Exchanger Materials in Simulated FBC Deposits", *Proc. NACE Conf. Corrosion-Erosion-Wear of Materials in Emerging Fossil Energy Systems* (Berkeley, CA), Jan 1982, p 493-510

12. Bakker, W., "Materials Performance in Coal Gasification Plants", *Proc. Conf. Corrosion-Erosion-Wear of Materials at Elevated Temperatures* (Berkeley, CA), Jan 1986, p 27-43

13. Rishel, D.M.; Pettit, F.S.; and Birks, N., Some Principal Mechanisms in the Simultaneous Erosion and Corrosion Attack of Metals at High Temperature, *Proc. NACE Conf. Corrosion-Erosion-Wear of Materials at Elevated Temperatures*, Paper No. 16 (Berkeley, CA), 1990

14. Gilmour, J.B., "The EMR-NSPC FBC Materials Tests," Report EPRI/PMRL 83-36 (OP), CANMET, Ottawa, Ontario, Canada, June 1983

15. Levy, A.V.; Slamovich, E.; and Jee, N., Elevated Temperature Combined Erosion-Corrosion of Steels, *Wear*, Vol 110 (No. 2), June 1986, p 117-149

16. Strangman, T.E., Thermal Barrier Coatings for Turbine Airfoils, *Thin Solid Films*, Vol 127, 1985, p 93-105

17. Sumner, I.E., and Ruckle, D., "Development of Improved Durability Plasma Sprayed Ceramic Coatings for Gas Turbine Engines," AIAA paper No. 80-1193, 1980

18. Levy, A.V., and Man, Y.F., Elevated Temperature Erosion-Corrosion of 9Cr1Mo Steel, *Proc. TMS-AIME Conf. High Temperature Corrosion in Energy Systems* (Detroit), Sept 1984, p 731-750

19. Levy, A.V., and Man, Y.F., "The Effect of Temperature on the Corrosion-Erosion of 9Cr1Mo Steel," *Wear*, Vol III (No. 2), Sept 1986, p 161-172

20. Lapides, L., and Levy, A.V., The Halo Effect in Jet Impingement Solid Particle Erosion Testing of Ductile Metals, *Wear*, Vol 58 (No. 2), Feb 1980, p 301-312

21. Dosanjh, S., and Humphrey, J., The Influence of Turbulence on Erosion by a Particle-Laden Fluid Jet, *Wear*, Vol 102 (No. 4), April 1985, p 309-330

22. Levy, A.V., and Man Y.F., "Nodular Scale Formation in Erosion-Corrosion of Iron-Chromium Alloys," Report No. LBL-19148, Lawrence Berkeley Laboratory, University of California, Berkeley, July 1985

23. Wood, G.C., and Whittle, DP, The Mechanism of Breakthrough of Protective Chromium Oxide Scales on Fe-Cr Alloys, *Corros. Sci. B.*, Vol 7, 1967, p 763

24. Boggs, WE, The Oxidation of Iron-Aluminum Alloys from 450 °C to 900 °C, *J. Electrochem Soc., Solid State SCI.*, June 1971, p 906

25. Skiyama, M.; Tomaszewiiz, P.; and Wallwork, G.R., Oxidation of Iron-Nickel Aluminum Alloys in Oxygen at 600 °C and 800 °C, *Oxid. Met.*, Vol 13 (No. 4), 1979, p 311

26. Kuroda, K.; Labun, P.A.; Welsch, G.; and Mitchell, T.E., Oxide Formation Characteristics in the Early Stages of Oxidation of Fe and Fe-Cr Alloys, *Oxid. Met.*, Vol 19, (No. 3/4), 1983

27. Levy, A.V., and Jee, N., "The Erosion-Corrosion of Scales on Heat Exchanger Alloys," Paper No. 475, NACE Corrosion 87 (San Francisco), 1987

28. Levy, A.V., and Man, Y.F., "Erosion-Corrosion Mechanisms and Rates in Fe-Cr Steels", *Wear*, Vol 131 (No. 1), May 1989, p 39-52

29. Levy, A.V., and Chik, P., The Effects of Erodent Composition and Shape on the Erosion of Steel, *Wear,* Vol 89 (No. 2), Aug 1986, p 151-162

30. Levy, A.V., Yan, J., and Patterson, J., Elevated Temperature Erosion of Steels, *Wear*, Vol 108 (No. 1), March 1986, p 43-60

31. Sorell, G., Elevated Temperature Erosion-Corrosion of Alloys in Sulfidizing Gas-Solid Streams: Parametric Studies, *Proc. NACE Conf. Corrosion-Erosion-Wear of Materials at Elevated Temperatures* (Berkeley, CA), 1986, p 204-229

32. Howes, M.A.H., Elevated Temperature Erosion-Corrosion of Alloys in Sulfidizing Gas/Solid Streams: Mechanistic Studies, *Proc. NACE Conf. Corrosion-Erosion-Wear of Materials at Elevated Temperatures* (Berkeley, CA), 1986, p 230-253

33. Natesan, K., Corrosion Behavior of Materials in Low and Medium BTU Coal Gasification Environments, *Proc. NACE Conf. Corrosion-Erosion-Wear of Materials in Emerging Fossil Energy Systems* (Berkeley, CA), 1982, p 100-136

34. Birks, N., Corrosion of High Temperature Alloys in Multicomponent Oxidative Environments, *Proc. Electrochemical Soc. Symp. Properties of High Temperature Alloys*, 1977

# Chapter 6
# Erosion-Corrosion of Materials in Elevated-Temperature Service

Any operating industrial system that utilizes fluid flows that contain small, solid particles is susceptible to combined erosion-corrosion (E-C). Such systems occur in many industries; among them are electricity generation, waste incineration, petroleum refining, chemical processing, cement manufacture and bulk transportation. Each system has its particular set of erodents, reactive fluids, and operating conditions. Because the basic mechanisms of erosion of ductile and brittle materials have distinct roots, understanding the erosion-corrosion that occurs in a particular type of service is helpful in addressing the material wastage that occurs in other services. However, the differences in the details of materials and operating conditions results in considerably different rates of loss, even though the basic mechanisms are the same, or quite similar.

One specific type of industrial service was selected as the basis for an in-depth study of erosion-corrosion; fluidized-bed combustion (Ref 1). Fluidized-bed combustors (FBC) are representative of a number of erosion-corrosion operating environments, consisting of an oxidizing gas at elevated temperature carrying erosive particles that are impacted against metallic or nonmetallic surfaces. In particular, the heat-exchanger tubes in FBCs were studied.

Successful simulation of the in-service operating conditions at the surface of a component such as a tube that is being degraded by erosion or erosion-corrosion requires that both the particle-laden fluid and its flow conditions against the target surface be reasonably well simulated. Because the exact conditions at the surface of an operating component are not precisely known, no matter whether they are steady state or transient, because of local chemical, physical, and fluid dynamics variations, the accuracy of the laboratory simulation has to be determined by an indirect method. This is done by directly comparing the material wastage rates and the chemical and morphological characteristics of the exposed surface of the same material exposed to laboratory and to in-service conditions. If they are acceptably near to each other, then the laboratory test is considered to be a useful simulation of the service environment.

The manner in which material wastage is measured in erosion-corrosion is an important consideration. It has been determined that the use of weight loss of a tested specimen is not a sufficiently accurate way to represent material wastage. When chemical reactive fluids and particles are directed at a test specimen, several different weight changes occur, over and above erosion-corrosion losses. The surfaces of the specimen, other than the erodent impacted area, can add or subtract weight from chemical reactions such as oxidation. The erodent particles can deposit or embed themselves on the test surface in quite significant layers. Mixtures of erodent particles and oxide scales can form on the eroding surfaces.

The most realistic manner in which to determine and report material wastage is to measure the actual thickness of the maximum loss area of the material before and after the test exposure, using devices accurate to a micron or less. The method for doing this uses a scribed grid placed on the ground surface to be eroded. The grid area thickness is measured at the grid intersections using a sensitive micrometer before testing. After testing the specimen is cross-sectioned and measured using a microscopic micrometer, relating the thickness at the location of the deepest penetration to the pretest grid thicknesses at the same position. This system has been shown to be reproducible to approximately 1 μm. It has been used to determine all of the data presented in this chapter.

## Erosion-Corrosion of Steel in Fluidized-Bed Combustors

The erosivity of FBC bed material particles is dependent upon a combination of the nature of the particles and the conditions under which they strike the surface of heat-exchanger tubes. These tubes are located in the bed, along the water walls, or in the convection pass of the FBC. Each area in which tubes are located has a different local, service environment (temperature, gas composition and particle velocity, impact angle, and solids loading), that directly relates to the metal wastage that the tubes experience. An equally important factor that determines the metal loss that occurs is the character of the bed material particles themselves (Ref 2-5). Their composi-

tion, shape, and size are major factors in establishing metal wastage rates of FBC heat-exchanger tubing materials. The bed material particle characteristics are strongly influenced by the nature of the feedstock material; that is, coal, wood chips, fruit pits, nuts, shells, limestone, sand, and what becomes of them in the combustion process.

The erosivity of various bed material and feedstock particles was determined in laboratory erosion-corrosion tests using combusted particles of various compositions, shapes, and sizes that were obtained from several commercially operating FBCs. The effects of individual characteristics of the particles on their erosivity were obtained in a standardized, laboratory, elevated-temperature erosion-corrosion test on 1018 plain carbon steel.

### Comparison of Laboratory Test and In-Service Behavior[(Ref 6)]

The laboratory test conditions selected simulated the exposure environment that occurs in circulating fluidizing bed combustors (CFBC) convection pass superheaters. How the laboratory test data compared to in-service metal wastage is shown in Table 6.1. The 1018 steel CFBC superheater tube metal loss should be compared to the three Lawrence Berkeley Laboratory (LBL) nozzle tester laboratory test results. It can be seen that the μm/h metal loss rate for all of the surfaces are in the same regime. The laboratory test loss rates listed were extrapolated from 5 h, steady-state loss tests; they agreed well with the 4000 h in-service exposure loss rates. Even the trends of the relatively small differences can be explained. For example, the lower loss rate in the

**Table 6.1  Comparison of metal wastage from laboratory and CFBC in service exposures**

| Exposure | Particle No. | Particle size, μm | Particle velocity, m/s | Test duration, h | Test temperature, °C | Target steel | Material loss, μm $\alpha = 30°$ | Material loss, μm $\alpha = 90°$ | Material loss rate, μm/h $\alpha = 30°$ | Material loss rate, μm/h $\alpha = 90°$ | Material loss rate, mils/1000 h $\alpha = 30°$ | Material loss rate, mils/1000 h $\alpha = 90°$ |
|---|---|---|---|---|---|---|---|---|---|---|---|---|
| CFBC primary superheater tube | 1 | 250 | 10 | 4000 | 538 | 1018 | 4686 | ... | 1.2 | ... | 47 | ... |
| CFBC secondary superheater tube | 1 | 250 | 10 | 4000 | 785 | 2.25Cr-1Mo | ... | 4166 | ... | 1.0 | ... | 42 |
| LBL tester | 1 | 250 | 20 | 5 | 450 | 1018 | 7.9 | 7.6 | 1.6 | 1.5 | 63 | 61 |
| LBL tester | 2 | 250 | 20 | 5 | 450 | 1018 | 6.6 | 5.4 | 1.3 | 1.1 | 53 | 43 |
| LBL tester | 2 | 250 | 10 | 8 | 450 | 1018 | 4.2 | ... | 0.5 | ... | 21 | ... |

10 m/s lab test in Table 6.1, 0.5 µm/h, compared to the higher, 1.2 µm/h, in-service tube loss was due to less-erosive, rounder-bed material particles being used in the laboratory test compared to more angular, erosive particles that were present in the CFBC. Both bed materials were from the same CFBC, but were from different runs using different feedstock.

Combined erosion-corrosion has been observed to reach steady state in a relatively short period of time over a wide range of conditions.

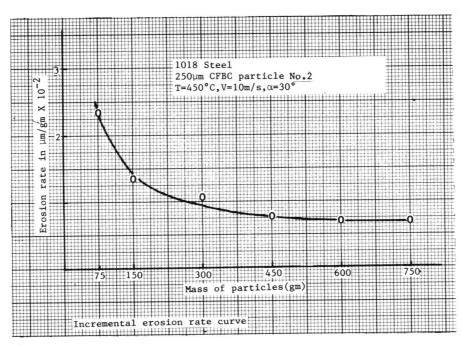

(a)

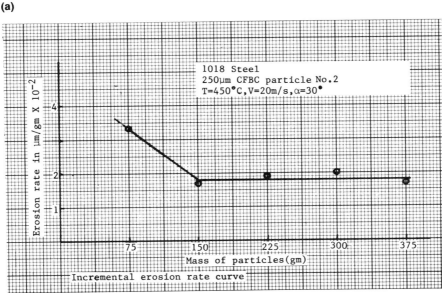

(b)

**Fig. 6.1** Incremental erosion-corrosion metal loss for $T = 450\ °C$. (a) $V = 10\ m/s$. (b) $V = 20\ m/s$

Figure 6.1 shows that it took 450 g of bed material particles at a flow rate of 1.25 g/min to reach steady-state conditions at $V = 10$ m/s and only 150 g at $V = 20$ m/s. Each data point in Fig. 6.1 is from a different specimen. Based on these curves, 600 g of erodent was used for all laboratory tester 10 m/s tests and 375 g for all 20 m/s tests.

Figure 6.2 shows sections from the U-bend of a 1018 primary superheater tube that had 4000 h of in-service exposure. Figure 6.3 is the surface and cross section of the area of the U-bend that had undergone the greatest material loss. A multiple-layer material was present on the tube surface after the 4000 h exposure that averaged 23 μm in thickness and consisted of a mixture of bed-material constituents (calcium, silicon,

aluminum) and iron oxide with the former decreasing and the latter increasing inward from the surface (see the EDS peak analyses in Fig. 6.3).

Figure 6.4 is the surface and cross section of the 1018 steel specimen from the 8 h, $V = 10$ m/s laboratory test listed in Table 6.1. A multiple-layer material was present on the surface that was similar in thickness to that on the heat-exchanger tube. Its outer layer was primarily bed material, while its inner layer was a mixture of bed material (calcium, silicon, aluminum) and iron oxide, as occurred on the in-service tube. It has been observed on both laboratory and in-service test surfaces that the outer layer can be either straight bed material or a mixture of it and iron oxide.

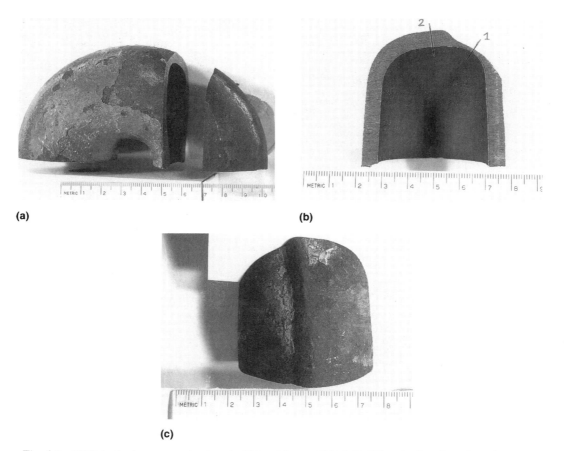

(a)

(b)

(c)

**Fig. 6.2** 1018 steel primary superheater tube U-bend from a CFBC. (a) U-bend at location of maximum metal wastage. (b) Cross section of U-bend at location of maximum metal wastage. (c) Outer surface of U-bend at location of maximum metal wastage

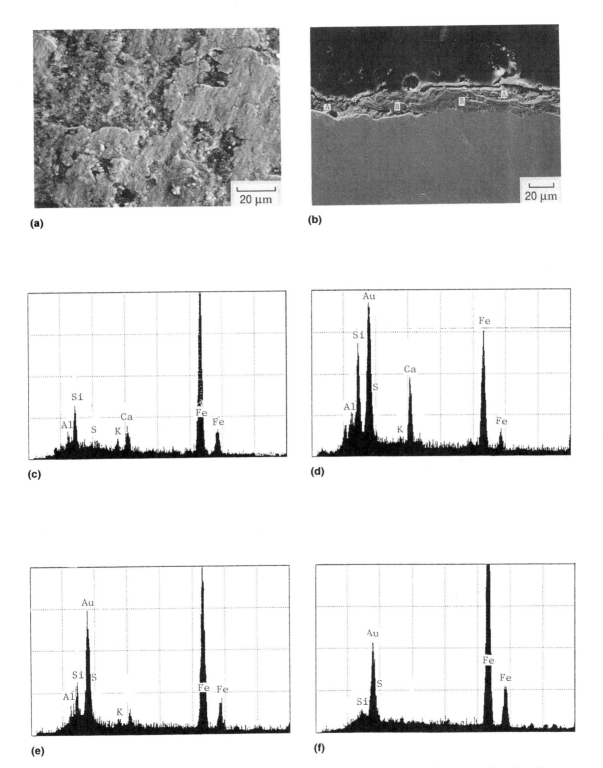

**Fig. 6.3** Surface and cross section of 1018 steel U-bend at area of greatest metal wastage (location 1). $T = 538\ °C$, $t = 4000$ h. (a) Surface. (b) Cross section. (c) Composition of surface. (d) Composition of position A. (e) composition of position B. (f) Composition of position C

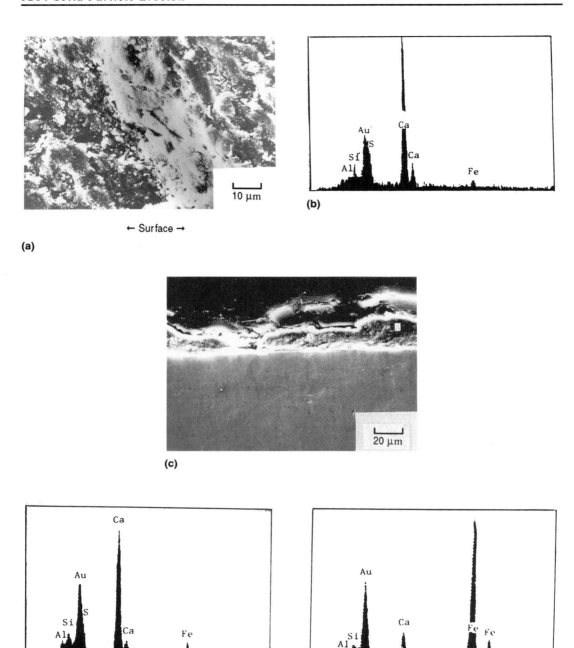

**Fig. 6.4** Surface and cross section of 1018 steel laboratory specimen (250 μm CFBC particle No. 2) $T = 450$ °C, $V = 10$ m/s, $t = 8$ h (600 g), $\alpha = 30°$. (a) Surface. (b) Composition of surface. (c) Cross section. Compositions of cross sections at: (d) position A and (e) position B

Figure 6.5 shows a laboratory test surface where both types of outer layers occurred, that is, bed material only and mixed-bed material and iron oxide. They were located at A and B, close together on the same surface indicating an intimate mixing of bed material and iron oxide. Also

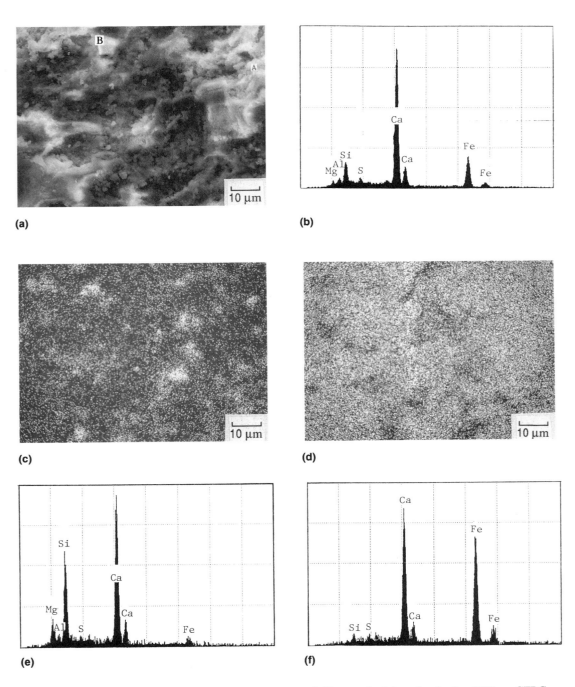

**Fig. 6.5** Surface of 1018 steel specimen with x-ray maps of silicon and calcium distribution (250 μm CFBC particle No. 2). $T = 450\ °C$, $t = 5$ h (375 g), $V = 20$ m/s, $\alpha = 30°$. (a) Surface. (b) Composition of surface. (c) Silicon map. (d) Calcium map. (e) Composition of position A. (f) Composition of position B

note in Fig. 6.5 how evenly the two major bed material constituents, calcium and silicon, were distributed over the surface.

Thus, erosion-corrosion occurs on both laboratory test specimens and CFBC convection pass superheater tube surfaces at approximately the same rate and in the same manner with a layer thickness that forms early on and remains essentially constant over thousands of hours of exposure (see Chapter 5).

### Test Conditions

The laboratory tests were carried out in the elevated-temperature nozzle tester (Chapter 5). Air was used as the carrier fluid for the particles, creating a generally oxidizing atmosphere, as occurs in FBCs. A standard set of test conditions was used to compare the erosivity of the various particles:

- Target metal, 1018 steel
- Temperature, $T = 450$ °C
- Velocity, $V = 20$ m/s
- Impact angle, $\alpha = 30°$
- Time, $t = 5$ h
- Solid loading = 375 g

These test conditions were selected to simulate as near as possible the erosion-corrosion environment in the convection pass, superheater region of a circulating fluidized bed combustor. Particle velocity was set in the tester using a computer program (Ref 7).

The tests used to determine the effects of various $SiO_2$-limestone mixtures such as occur in

FBCs were carried out at a temperature of 450 °C, an impact angle of 30°, a particle velocity of 20 m/s for $SiO_2$-limestone mixtures tests, particle velocities of 10, 20, and 30 m/s for $SiO_2$-calcined limestone mixtures, and particle velocities of 2.5 and 20 m/s for $SiO_2$-calcium sulfate mixtures. Three test times were used at three particle loadings; 5 h for 375 g (at $V = 20, 30$ m/s), 8 h for 600 g (at $V = 10$ m/s), and 100 h for 7500 g (at $V = 2.5$ m/s).

### Materials

The target material for all of the erosion-corrosion tests was 1018, plain carbon steel, which is commonly used in bubbling fluidized bed combustors (BFBC) for in-bed evaporator tubes, circulating fluidized bed combustors (CFBC) for water-wall tubes and convection-pass primary superheater tubes, and in pressurized fluidized bed combustors (PFBC). Sixteen different erodent particles from ten operating FBCs as well as five different sand and limestone particles used in two of the FBCs were tested. Several of the as-received bed materials are shown in Fig. 6.6, and sand and limestone particles are shown in Fig. 6.7. The sixteen different bed materials have been designated No. 1 through 16. Their sources (Units A through J), size and type (BFBC, CFBC, and PFBC) are listed in Table 6.2. The particles

**Table 6.2  Source, size, and type of bed material and metal wastage**

| Erodent particle No. | Type | Unit | Average size, μm diam | Thickness loss, μm |
|---|---|---|---|---|
| 1 | CFBC | A | 250 | 7.9 |
| 2 | CFBC | A | 250 | 6.6 |
| 3 | CFBC | C | 365 | 5.1 |
| 4 | CFBC | B | 300 | 5.3 |
| 5 | BFBC (0.03% Cl) | D | 250 | 3.2 |
| 6 | BFBC (0.3% Cl) | D | 250 | 2.3 |
| 7 | BFBC | E | 250 | 3.1 |
| 8 | CFBC | A | 250 | 4.0 |
| 9 | CFBC | F | 250 | 2.8 |
| 10 | CFBC | G | 250 | 3.9 |
| 11 | BFBC | H | 770 | 2.1 |
| 12 | PFBC | I | 670 | 6.7 |
| 13 | BFBC | J | 570 | 3.4 |
| 14 | BFBC | J | 720 | 5.8 |
| 15 | BFBC | D | 800 | 2.6 |
| 16 | CFBC | A | 370 | 4.7 |
| Angular sand | | A | 450 | 12.5 |
| Round sand | | A | 400 | 3.3 |
| Angular sand | | B | 230 | 5.5 |
| Limestone | | A | 800 | 5.8 |
| Limestone | | B | 300 | 4.5 |

Note: Test conditions: material, 1018 steel, $T = 450$ °C, $\alpha = 30°$, $V = 20$ m/s, $t = 5$ h (375 g loading), air

from the CFBCs were taken at a location in the convection pass below the super-heater tube banks. It is not known from where the BFBC and PFBC particles were taken, but it is assumed that they were taken from bed drains.

The particles have a wide range of particle sizes with those from the BFBCs having a much wider size range than those from the CFBCs, including particles considerably larger than 1 mm in diameter. To be able to compare the erosivity of different bed-material particles, an average particle size of 250 μm was screened out and used as a standard test size. By screening the actual bed material to obtain one size of particles, an un-

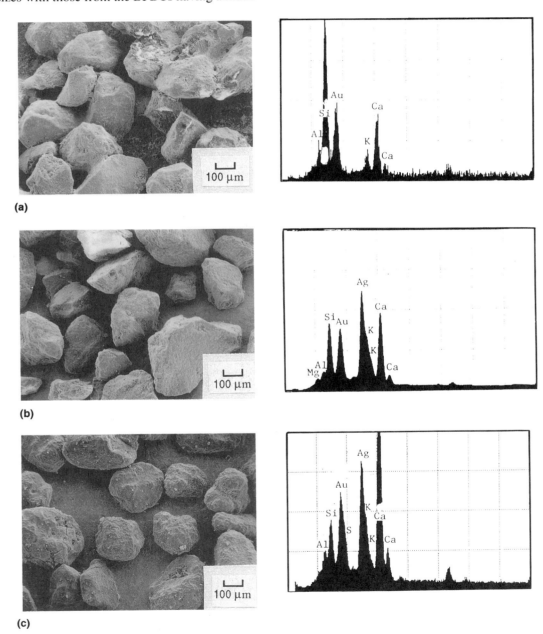

**Fig. 6.6**   Screened 250 μm CFBC bed material particles. Micrographs and compositions: (a) Particle No. 1. (b) Particle No. 2. (c) Particle No. 4

**Fig. 6.7** Screened CFBC sand and limestone particles. (a) Unit A sand. (b) Unit B sand. (c) Round sand (unit A). (d) Limestone (unit A). (e) Limestone (unit B)

known degree of particle selection occurred, as the erosion characteristics of larger and smaller particles in a total bed material can differ somewhat. To account for this, the actual, as-received, particle-size mixtures from the different bed materials were also used directly as erodent particles.

The $SiO_2$-$CaCO_3$, $SiO_2$-$CaO$ and $SiO_2$-$CaSO_4$ mixtures were prepared in the laboratory from feedstock of CFBC Unit B and reagent grade $CaSO_4$. Limestone was calcined in a static air furnace at 950 °C for 12 h. Figure 6.8 shows the appearance of the $SiO_2$, limestone and cal-

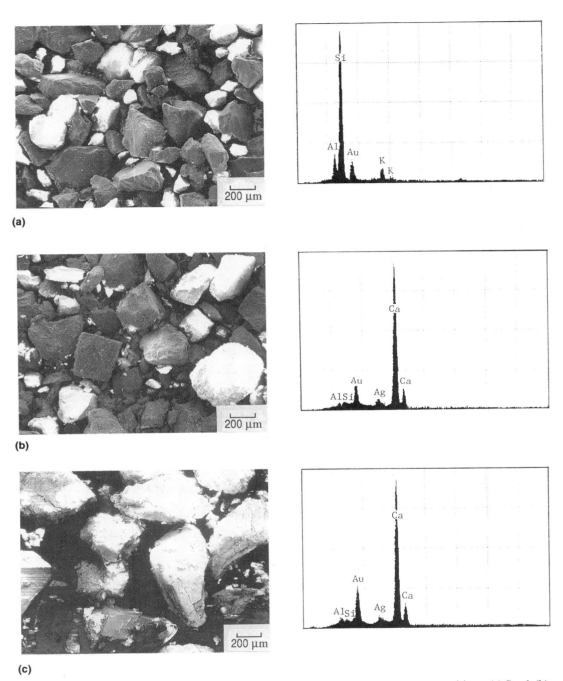

**Fig. 6.8** Sand and limestone particles used to prepare mixtures. Micrographs and compositions: (a) Sand. (b) Limestone. (c) Calcined limestone

cined limestone used in the mixtures. The white-appearing particles are the result of charging in the scanning electron microscope (a gold flash was used to minimize this). The EDS silver peak is from the adhesive used to bond the particles to the pedestal of the microscope. An average particle size of 250 μm was used in the mixture tests. The particles were generally quite angular in shape.

## Erosion-Corrosion Rates[Ref 8]

The metal wastage of specimens eroded-corroded by the different erodent particles are listed in order of erosivity ranking in Table 6.3. The particles that caused the greatest metal loss in the 5 h tests had the most angular shape (Ref 9), the highest amount of $SiO_2$, and the lowest amount of CaO or $CaSO_4$. The effects of these characteristics will be discussed. The results from these test series have a significant impact on the makeup of the feedstocks used in FBCs because of the influence the feedstocks and their resultant bed materials have on the surface degradation rates of combustor components.

When the as-received size particles were used as erodents, generally the larger particles were more erosive than the smaller ones. However, this effect varied considerably, as can be seen in Table 6.3. The smaller CFBC particles were generally more erosive than the larger BFBC particles. The largest particle, No. 15, at 800 μm in diameter, had a lower erosivity than several of the 250 μm diam bed material particles because of its other characteristics. Where the same type of particle was tested at two different sizes; for example, particles No. 13 and 14, the larger particle was more erosive, although other characteristics also contributed to the difference. On the other hand, an increase in the size of particle No. 11 from 250 to 770 μm caused almost no change in its erosivity. Also, the same-size particle from the same FBC, particles No. 1 and 2, had different erosivities because of their other characteristics. In Table 6.2, it can be seen that round $SiO_2$ was much less erosive than angular $SiO_2$ at nearly the same particle size because of the difference in their shapes.

Table 6.3 shows that, with one exception, CFBC bed materials had higher erosivity than BFBC or PFBC bed material particles. This difference was primarily due to the fact that the CFBC bed materials had a higher $SiO_2$ content as the result of purposeful $SiO_2$ additions to the combustor feedstock. The BFBC and PFBC only contained $SiO_2$ from the coal fuel. The reason for the one exception (CFBC particle No. 9) is related to particle shape and composition, as will be discussed later. In Table 6.3 it can also be seen that among the different CFBC bed materials, particles No. 1 and 2 had the highest erosivity. Particles No. 8 and 16 had much lower erosivity than particles No. 1 and 2 even though they were all from CFBC unit A. BFBC bed material particles No. 13, 15, and 11 had the lowest erosivity of all of the particles tested.

### Surface Layer Characteristics

**CFBC.** The microstructures of the surfaces of the erosion-corrosion specimens correlated with their metal-loss rates. As the layer on the target

**Table 6.3  Ranking of erosivity of FBC bed material particles**

| Bed material particle No. | Unit | Type | Metal loss, μm | | Erosivity ranking |
| | | | 250 μm particle size | Other particle size | |
|---|---|---|---|---|---|
| 1 | A | CFBC | 7.9 | ... | 1 |
| 2 | A | CFBC | 6.6 | ... | 2 |
| 4 | B | CFBC | 5.4 | ... | 3 |
| 3 | C | CFBC | 4.8 | 5.1 (365 μm) | 4 |
| 8 | A | CFBC | 4.0 | ... | 5 |
| 16 | A | CFBC | 3.95 | 4.7 (370 μm) | 6 |
| 10 | G | CFBC | 3.9 | ... | 7 |
| 5 | D | BFBC | 3.2 | ... | 8 |
| 7 | E | EFBC | 3.1 | ... | 9 |
| 9 | F | CFBC | 2.8 | ... | 10 |
| 12 | I | PFBC | 2.5 | 6.7 (670 μm) | 11 |
| 6 | D | EFBC | 2.3 | ... | 12 |
| 14 | J | BFBC | 2.3 | 5.8 (720 μm) | 12 |
| 13 | J | BFBC | 1.9 | 3.4 (570 μm) | 13 |
| 15 | D | BFBC | ... | 2.6 (800 μm) | 14 |
| 11 | H | BFBC | 1.9 | 2.1 (770 μm) | 15 |

Note: Test conditions: material, 1018 steel, $T = 450$ °C, $\alpha = 30°$, $V = 20$ m/s, $T = 5$ h (375 g loading), air

surfaces changed character, that is, thickness, composition, morphology, and continuity, the substrate metal wastage rate changed. Figure 6.9 shows the thin, intermittent layer that formed on the 1018 steel as the result of impacts by the angular, most erosive particle, No. 1, which came from wood chip burning unit A. From the EDS peak analyses, it can be seen that there was a low calcium content and a relatively high silicon content in the mixed iron oxide-bed material scale segments. When the bed material consisted of rounder particles with higher $CaSO_4$, CaO contents (particle No. 4), the metal loss in the 5 h laboratory test decreased (see Table 6.3).

The principal phases contained in the erodent particles are listed in Table 6.4. The highest content phase is listed first. Other minor constituents in the bed materials ( wt %) include $Al_2O_3$, $K_2O$, MgO, $TiO_2$, $Fe_2O_3$. Table 6.4 also lists the microhardness of the principal constituents of the bed materials, $SiO_2$, and the calcium compounds (which were identified by SEM calcium x-ray maps).

The surface and cross section of the scale layer on the 1018 steel impacted by particle No. 4 are shown in Fig. 6.10. The cross section was similar in thickness and morphology to that caused by the somewhat more erosive particle No. 2 from Unit A. However, because the coal fuel used in Unit B contained sulfur, a greater amount of limestone was therefore used in its feedstock than in Unit A; the principal constituent of the outer part of the scale layer was $CaSO_4$ (see Table 6.4). The

$CaSO_4$ and the CaO present acted as a cement to bond the layer together and make it more protective. The rounder shape of the particles also contributed to the lower metal loss.

The least erosive particle from Unit A was particle No. 8. The sand used in this run was round Ottawa sand, which was much less erosive than the more angular-shaped sands used when particles No. 1, 2, and 4 were generated in Unit A.

The least erosive bed material particles from a CFBC, particle No. 9, are shown in Fig. 6.11(a). Table 6.3 shows that they caused a metal loss that was 65% less than that caused by particle No. 1. Figure 6.11(a) shows that the particles were quite round, the particle characteristic that had the greatest effect on its erosivity (Ref 3). The morphology of the scale layer was continuous and had high $CaSO_4$, CaO content throughout (Table 6.4). It contained more calcium near the scale-metal interface than the layers formed by other CFBC particles, as can be seen in the EDS peaks (Fig. 6.11g). All of these positive factors contributed to the low erosivity. The CFBC from which these particles came is known to have a benign environment and has experienced minimal tube wastage in long-time service.

The presence of the calcium compound cementing agent in the scale changes its erosion behavior from that described in Chapter 5 for continuous scales. The calcium-bonded scale is considerably stronger than the silica-iron oxide

**Table 6.4  Microhardness and phases of FBC particles**

| Erodent particle No. (unit No.) | Microhardness (100 HV) (from 10 determinations) | | X-Ray diffraction analyses, phase (larger quantities listed first) |
|---|---|---|---|
| | $SiO_2$ | CaO/$CaSO_4$, $CaCO_3$ | |
| No. 1 (Unit A) | 1232 | 112 | $SiO_2$, CaO |
| No. 2 (Unit A) | 881 | 110 | $SiO_2$, CaO |
| No. 3 (Unit C) | 1074 | 59 | $SiO_2$, CaO, $CaCO_3$ |
| No. 4 (Unit B) | 878 | 98 | $CaSO_4$ (more), $SiO_2$, CaO, $Al_2O_3$, $CaCO_3$ |
| No. 5 (Unit D) | 470 | 99 | $CaSO_4$, CaO, $SiO_2$, $Al_2O_3$ |
| No. 7 (Unit E) | 777 | 99 | $CaSO_4$, CaO, $SiO_2$ |
| No. 8 (Unit A) | 1067 | 63, 331 | CaO, $CaCO_3$, $SiO_2$ |
| No. 9 (Unit F) | 589 | 114, 208 | $CaSO_4$, CaO, $CaCO_3$, $SiO_2$, MgO |
| No. 11 (Unit H) | ... | 138, 251 | $CaSO_4$, CaO, $CaCO_3$, $SiO_2$ |
| No. 15 (Unit D) | 545 | 166, 236 | $CaSO_4$, $SiO_2$, $Al_2O_3$, MgO |
| Sand (Unit A) | 1329 | ... | $SiO_2$ |
| Sand (Unit B) | 981 | ... | $SiO_2$ |
| Round sand (Unit A) | 1273 | ... | $SiO_2$ |
| Limestone (Unit A) | ... | 288 | $CaCO_3$ |
| Limestone (Unit B) | ... | 192 | $CaCO_3$ |

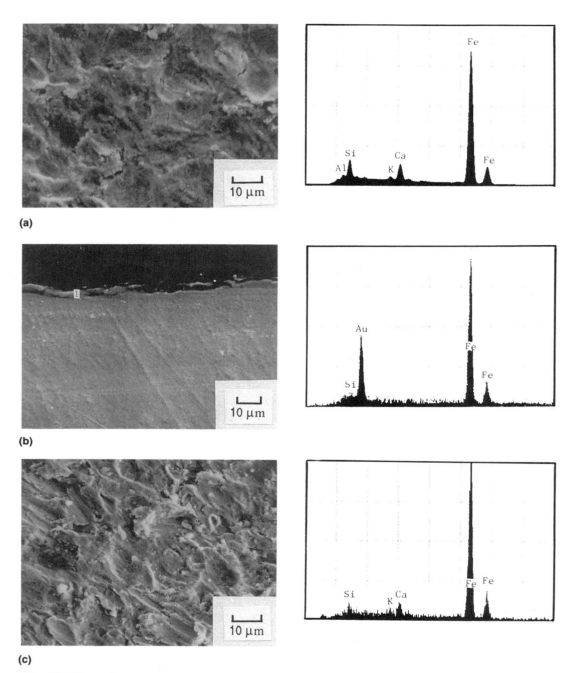

**Fig. 6.9** The surface, cross section, and the EDS peak analyses of 1018 steel eroded-corroded by 250 μm CFBC particle No. 1. $V = 20$ m/s, $\alpha = 30°$, $t = 5$ h (375 g). Micrographs and compositions: (a) Surface ($T = 450$ °C). (b) Cross section ($T = 450$ °C). (c) Surface ($T = 25$ °C)

scale described in Chapter 5 and that which occurred in the scales from particles No. 1 and 2; it is much more protective. Compare the EDS peaks in Fig. 6.9 and Fig. 6.11. The scale from particle No. 9 has high calcium peaks, while that from particle No. 1 has a very small calcium peak. The weak $SiO_2$-$Fe_2O_3$ scales are more prone to spalling (see Fig. 5.14 and Fig. 6.9). The strength of the eroding scale is a greater factor than morphology in erosion-corrosion.

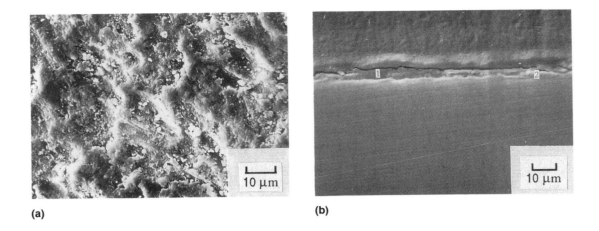

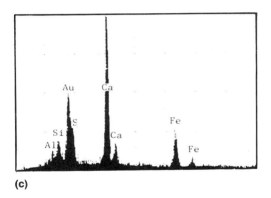

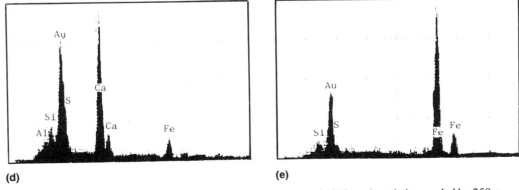

**Fig. 6.10** The surface, cross section, and the EDS peak analyses of 1018 steel eroded-corroded by 250 μm CFBC particle No. 4. (a) Surface. (b) Cross section. (c) Composition of surface. (d) Composition of cross section (position 1). (e) Composition of cross section (position 2)

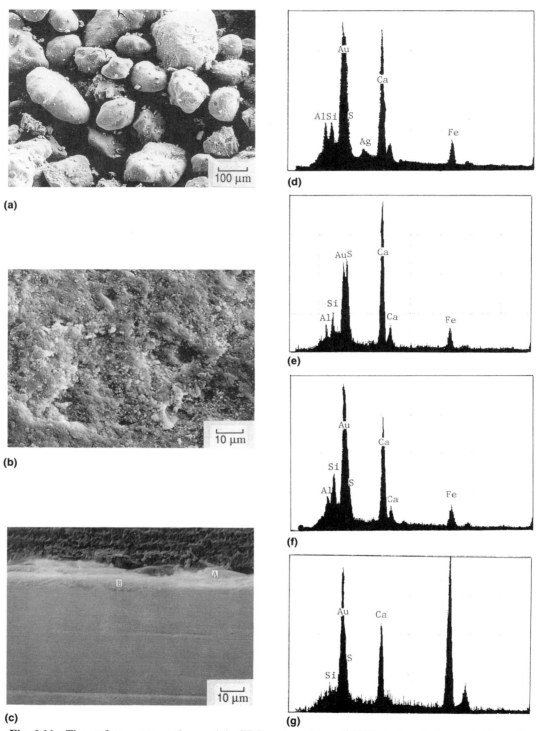

**Fig. 6.11** The surface, cross section, and the EDS peak analyses of 1018 steel eroded-corroded by 250 μm CFBC particle No. 9 and the appearance of the particles. $T = 450$ °C, $V = 20$ m/s, $t = 5$ h (375 g), $\alpha = 30$°. (a) CFBC particle No. 9 (unit F). (b) Surface. (c) Cross section. (d) Composition of particle No. 9 (unit F). (e) Composition of surface. (f) Composition of cross section (position A). (g) Composition of cross section (position B)

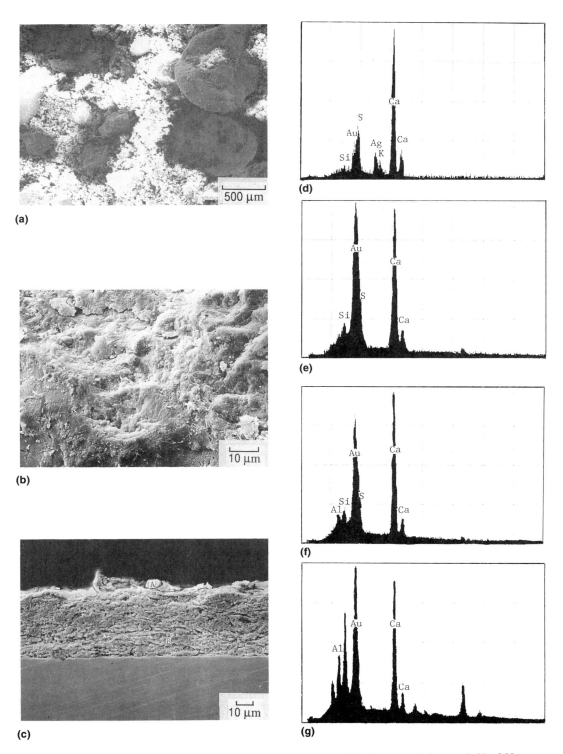

**Fig. 6.12** The surface, cross section, and the EDS peak analyses of 1018 steel eroded-corroded by 250 μm BFBC particle No. 13. $T = 450$ °C, $V = 20$ m/s, $t = 5$ h (375 g), $\alpha = 30°$. (a) Particle No. 13 (unit J). (b) Surface. (c) Cross section. (d) Composition of particle No. 13. (e) Composition of surface. (f) Composition of cross section (position A). (g) Composition of cross section (position B)

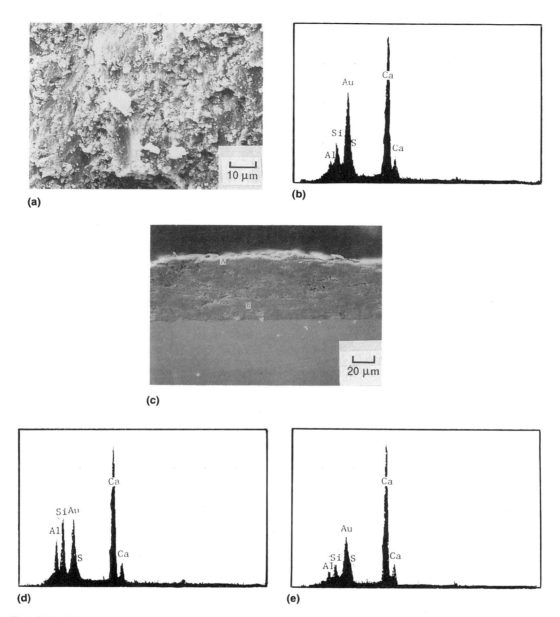

**Fig. 6.13** The surface, cross section, and the EDS peak analyses of 1018 steel eroded-corroded by 250 μm BFBC particle No. 14 (Unit J). $T$ = 450 °C, $V$ = 20 m/s, $t$ = 5 h (375 g), α = 30°. (a) Surface. (b) Composition of surface. (c) Cross section. (d) Composition of cross section (position A). (e) Composition of cross section (position B)

**BFBC**. The particles that came from BFBCs formed thicker surface layers in the standard laboratory test than those from CFBCs. Figure 6.12 shows the layer from BFBC particle No. 13, which was quite round in shape. As in all other tests, the scale layer developed early in the test and then maintained a constant thickness. The layer was typical of those formed by BFBC particles by having only bed material with essentially no iron oxide in it until quite near to the scale metal interface, as seen in the EDS peaks (Fig. 6.12g). The 1.9 μm metal loss listed in Table 6.3 was the lowest for any particle tested.

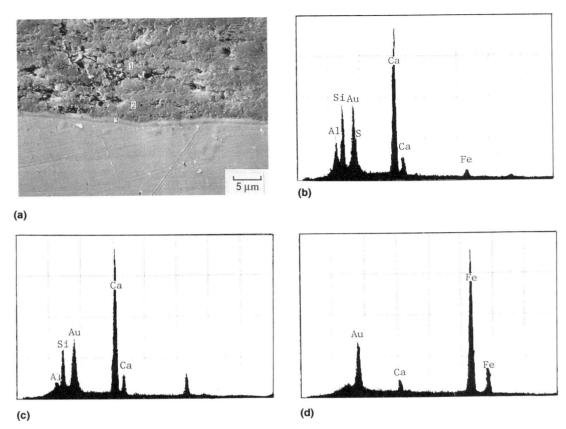

**Fig. 6.14** Interface area of 1018 steel eroded-corroded by particle No. 14 (Unit J). $T = 450\,°C$, $V = 20$ m/s, $t = 5$ h (375 g), $\alpha = 30°$. (a) Cross section. (b) Composition of cross section (position 1). (c) Composition of cross section (position 2). (d) Composition of cross section (position 3)

BFBC particle No. 14 caused the surface layer to form on the 1018 steel that is shown in Fig. 6.13 when the 250 µm diam particle size was used in the standard test. It is somewhat layered and contains only bed material down to very near the substrate metal. It is only when the interface area is viewed at a higher magnification in Fig. 6.14 that the presence of the iron oxide layer can be discerned. A thin, continuous, iron oxide scale with a small amount of calcium in it occurred near the scale-metal interface.

### Scale-Deposit Layer Composition Distribution

The distribution of the constituents through the scale layer can be seen in the surface EDS peaks and x-ray maps in Fig. 6.5 and from line scan analyses and x-ray maps of the cross sections in Fig. 6.15 and 6.16. Figure 6.15 shows that in the outer part of the scale layer eroded-corroded by a high silicon content CFBC bed material, there was a mixture of bed material containing calcium, silicon, and aluminum compounds and iron oxide (determined by x-ray diffraction). Nearer the scale-metal interface, the bed material content of the layer was markedly decreased and iron oxide was primarily present. The typical scale layer formed by BFBC particles containing higher calcium content had only bed material with essentially no iron oxide in it until quite near to the scale-metal interface, as seen in Fig. 6.16.

From the iron and calcium line analyses of the cross sections shown in Figures 6.15 and 6.16, it can be seen that no matter whether CFBC particles or BFBC particles were used for the erodent, the distribution of iron and calcium had linear gradients in the inner layer,

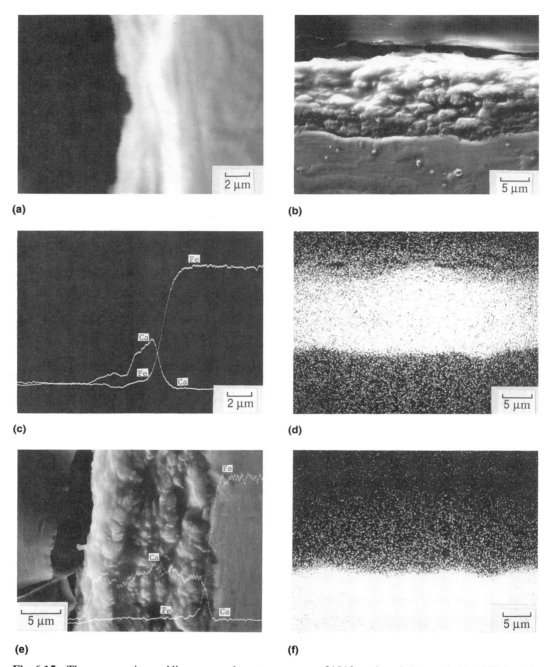

**Fig. 6.15** The cross section and line scan analyses, x-ray maps of 1018 steel eroded-corroded by CFBC particles, No. 4 and 8. $T = 450\ °C$, $V = 20\ m/s$, $t = 5\ h$ (375 g), $\alpha = 30°$. (a) Cross section, particle No. 4. (b) Cross section, particle No. 8. (c) Line scan, particle No. 4. (d) Calcium map, cross section, particle No. 8. (e) Cross section and line scan, particle No. 8. (f) Iron map, cross section, particle No. 8

near the interface. The composition gradient of iron reduced in the outward direction and that of calcium reduced in the inward direction. Oxygen is thought to vary through the scale in the same way that calcium does. This indicates that cation diffusion of iron atoms occurred outward and

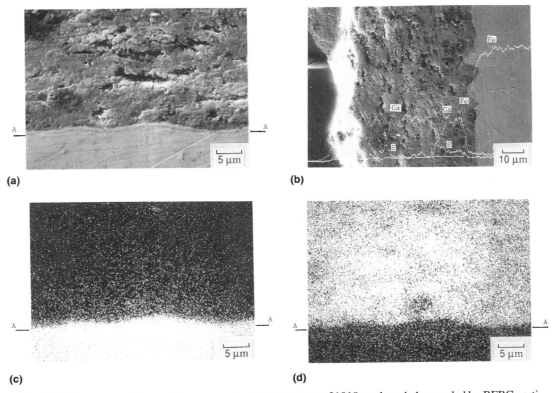

**Fig. 6.16** The cross section and line scan analyses, x-ray maps of 1018 steel eroded-corroded by BFBC particle No. 14. $T = 450$ °C, $V = 20$ m/s, $t = 5$ h (375 g), $\alpha = 30°$. (a) Cross section. (b) Cross section and line scan. (c) Iron map, cross section. (d) Calcium map, cross section

anion diffusion of oxygen occurred inward to form oxide and somehow combine with calcium. Thus, in the inner layer near the interface, a chemical diffusion process occurred.

The distribution of calcium and sulfur in the outer layer is neither linear nor can it be directly correlated with the distance from the outside surface to the scale layer-metal interface. This means that in the outer, thicker layer the calcium and sulfur compounds in the bed material deposited on the surface of the samples and were mechanically mixed in with the iron oxide that was forming from the base metal. The intimacy of this mixing can be seen in Fig. 6.5. The distance between positions A and B in the micrograph of the surface is a few microns. Yet position A has essentially no iron oxide, while position B has significant iron oxide. Therefore, the composition distribution of iron oxide and bed material in the scale layer depends on both chemical and

mechanical processes. In the inner layer, the chemical process predominated, while in the outer layer the mechanical process primarily occurred.

### Erodent Compositions[Ref 10]

**SiO$_2$-Limestone (CaCO$_3$) Mixtures.** The relationship between the erosive SiO$_2$ and the softer calcium compounds that became a part of the scale layer on eroded-corroded steel surfaces was investigated by making a series of mechanical mixtures of SiO$_2$ and limestone (CaCO$_3$), SiO$_2$ and calcined limestone (CaCO), and SiO$_2$ and calcium sulfate (CaSO$_4$) ranging from 100% SiO$_2$ to 100% calcium compound and determining the erosivity of the resulting mixtures. Figure 6.17 shows the curve of the metal thickness loss versus the proportions of SiO$_2$ to CaCO$_3$. It can be seen that at the higher proportions of SiO$_2$ to CaCO$_3$, the metal thickness losses were greater

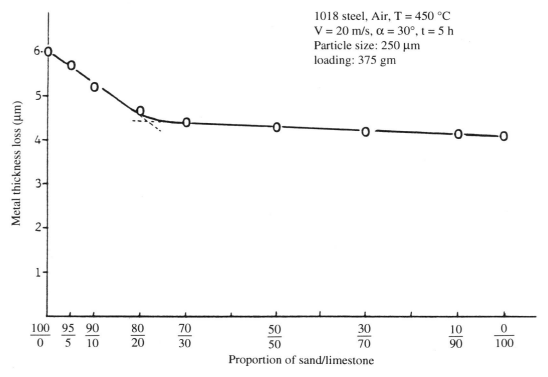

1018 steel, Air, T = 450 °C
V = 20 m/s, α = 30°, t = 5 h
Particle size: 250 μm
loading: 375 gm

**Fig. 6.17** Metal thickness loss versus $SiO_2$/limestone ($CaCO_3$) erodent mixtures

and increased with increasing amounts of $SiO_2$. A sharp transition occurred at the 80/20 ratio, on either side of which the curve was linear. The higher $CaCO_3$ content erodents caused lower metal thickness losses. The slope of the curve was low for the lower $SiO_2$ content mixtures compared to the higher $SiO_2$ content mixtures.

Figure 6.18 shows the cross section of specimens eroded-corroded by different $SiO_2/CaCO_3$ mixtures, along with their EDS analyses. It can be seen that when the 95/5 $SiO_2/CaCO_3$ mixture was used, no protective layer containing calcium formed on the surface. The metal was directly deformed and eroded by the impacting particles. Some evidence of metal platelet formation can be seen.

At the transition ratio of 80/20, a protective layer that contained some calcium had formed on the metal. Beyond the transition ratio of 80/20, a protective layer with a high calcium content formed (Fig. 6.18e), which resulted in lower metal wastage. Combining Fig. 6.17 with Fig. 6.18 it

can be seen that the transition ratio is related to the formation or lack of formation of a protective layer on the metal surface. Once there was enough calcium in the erodent to form a protective layer, further increases in the amount of limestone had only a minor effect on the amount of metal loss. The small but continued decrease in the metal loss in this region of the curve with greater $CaCO_3$ content resulted from the decreasing amount of $SiO_2$ that was in the erodent mixtures.

**Sand-Calcined Limestone Mixtures.** Figure 6.19 shows curves of the metal thickness loss versus the proportions of $SiO_2/CaO$ at three particle velocities. It can be seen that when using $SiO_2/CaO$ mixtures instead of $SiO_2/CaCO_3$ mixtures for the erodent, the same shape curves occurred with the higher percentages of $SiO_2/CaO$ causing greater metal losses. A transition ratio also occurred, on both sides of which the metal losses were linear. Moreover, at the transition ratio the metal loss is essentially the same, 2.3 μm, at all three particle velocities.

**(a)**

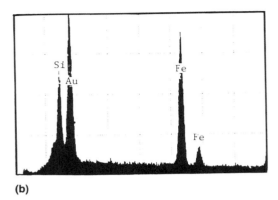

**(b)**

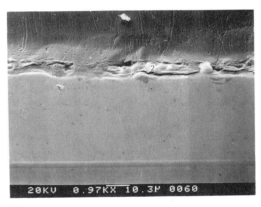

**(c)**

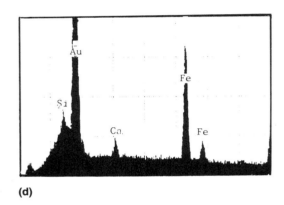

**(d)**

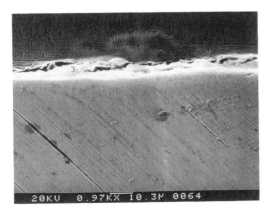

**(e)**

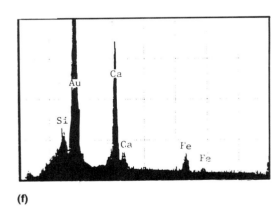

**(f)**

**Fig. 6.18**   Cross sections of 1018 steel surfaces eroded-corroded by $SiO_2/CaCO_3$ erodent mixtures. (a) $SiO_2/CaCO_3$: 95/5. (b) Composition of cross section (position 1). (c) $SiO_2/CaCO_3$: 80/20. (d) Composition of cross section (position 2). (e) $SiO_2/CaCO_3$: 70/30. (f) Composition of cross section (position 3)

More CaO is required to reach the transition ratio as the particle velocity increases. This is due to the greater force of the impacting particles at the higher velocities requiring more calcium compound to be able to initiate the formation of a protective layer. For a given CaO content, metal loss increased with particle velocity. Beyond the transition ratio, at higher CaO contents, the in-

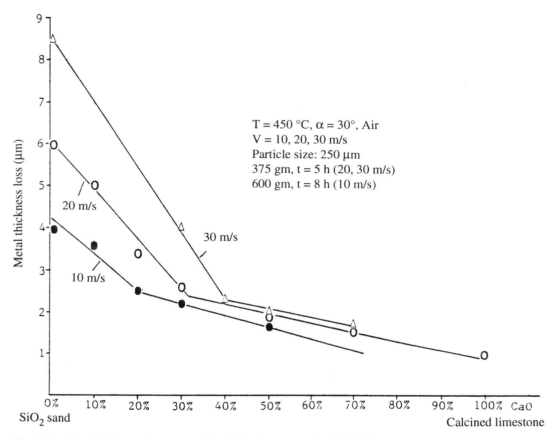

**Fig. 6.19** Metal thickness loss versus $SiO_2/CaO$ mixtures at $V = 10, 20, 30$ m/s

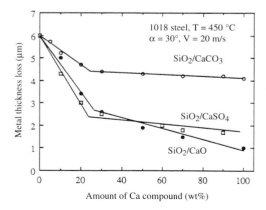

**Fig. 6.20** Metal thickness loss versus $SiO_2/Ca$ compound mixtures

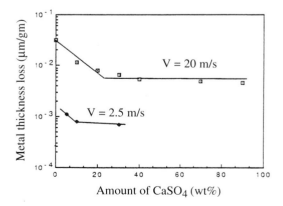

**Fig. 6.21** Metal thickness loss rate versus amount of $CaSO_4$ at 20 and 2.5 m/s particle velocities

crease was quite small and the metal loss became a function of the erosivity of the particle rather than the particle velocity. Prior to the transition ratio, at higher $SiO_2$ contents, the metal wastage had a strong dependency on particle velocity.

The difference in the effect of particle velocity on metal wastage strongly related to the formation of the protective layer. At higher, angular-shaped $SiO_2$ contents, the metal itself was participating in the high metal wastage, elevated-temperature erosion process. When the CaO content exceeded that of the transition ratio, a protective layer formed that prevented direct participation by the base metal in the metal-loss process. The protective layer was being eroded and the metal-wastage rate decreased. The loss of the layer is a function of the erosivity of the erodent, which slowly decreased with decreasing $SiO_2$ content. Comparing Fig. 6.17 and Fig. 6.19, the calcined limestone particles can be seen to have a lower erosivity than the limestone particles. At a CaO particle velocity of 20 m/s, the metal thickness loss was only 1 μm, which is one-fourth that caused by $CaCO_3$ (4.1 μm).

**$SiO_2$-Calcium Sulfate Mixtures.** Figure 6.20 shows the curve of the metal thickness loss versus the amount of $SiO_2$ in $SiO_2$/Ca compound mixtures (Ref 11). The transition ratio for the $SiO_2$/$CaSO_4$ mixtures occurred at nearly the same ratio as the $CaCO_3$ and CaO containing mixtures, almost a 75/25 ratio of $SiO_2$/$CaSO_4$. Beyond the transition ratio, at higher $CaSO_4$ contents, a dense, adherent, protective deposit layer formed on the surface of the steel. The deposit has two deterrent effects on erosion-corrosion. It prevents the erodent particle from directly impinging on the metal surface, and it prevents the metal surface from being directly exposed to the oxygen atmosphere. Scanning electron microscopy and x-ray diffraction analysis of the tested surfaces determined that no iron sulfide (FeS) occurred in the calcium sulfate deposits, indicating that sulfidation did not occur.

**Effect of Particle Velocity.** Figure 6.21 shows two curves of metal thickness loss rates versus the amount of calcium sulfate at 20 and 2.5 m/s particle velocities. It can be seen that when the particle velocity was 2.5 m/s, the shape of the curve was the same as those in Fig. 6.20. A transition ratio also occurred, but at 90% $SiO_2$ rather than 75% $SiO_2$ as was the case at the 20 m/s velocity. At the higher particle velocity, the deposit on the surface required more calcium sulfate to be able to initiate the formation of the protective layer, as occurred in the $SiO_2$/CaO tests (see Fig. 6.19). The transition ratio increased as the particle velocity decreased, with less calcium compound being required for the formation of a protective layer at the lower velocity.

Figure 6.22 and 6.23 show the morphological and composition features of the cross sections of the scale layers on erosion-corrosion specimens impacted by erodent mixtures containing 90 and 70% $SiO_2$ at the lower particle velocity of 2.5 m/s, along with EDS peak analyses. It can be seen that a relatively thick, continuous scale formed on the surface of the specimens. Heavy electronic charging occurred on the surface layers. Figures 6.21 to 6.23 show that the transition ratio is related to the formation or lack of formation of a $CaSO_4$ deposit/scale protective layer on the erosion-corrosion surface. The iron peak increased the nearer to the scale layer-metal interface the EDS analysis was made. This indicates that a significant outward transport of iron occurred from the substrate into the $CaSO_4$ deposit, where it was oxidized. This behavior has been observed in all of the work where bed materials were used

**Table 6.5   X-ray diffraction analyses of scale layers**

| Sample No. | Composition of mixture | Phases, wt% | | | | | |
|---|---|---|---|---|---|---|---|
| | | Fe | $Fe_2O_3$ | $Fe_3O_4$ | $CaSO_4$ | $SiO_2$ | FeS |
| C31 | 95/5 $SiO_2$/$CaSO_4$ | 1.6 | 8.9 | 60.1 | 27.7 | 1.7 | ... |
| C30 | 90/10 $SiO_2$/$CaSO_4$ | 5.0 | 19.1 | 10.1 | 63.5 | ... | 2.3 |
| C33 | 70/30 $SiO_2$/$CaSO_4$ | 2.0 | 6.8 | 10.8 | 76.6 | ... | 3.8 |

Note: Test conditions: $T = 450 \, °C$, $V = 2.5$ m/s, $\alpha = 30°$, $t = 100$ h as the erodent at test temperatures where iron oxide could form (Ref 12).

Combining the EDS peak analyses in Fig. 6.23 with the x-ray diffraction analysis results listed in Table 6.5, it can be seen that an FeS phase existed at the deposit/scale layer-substrate interface of the specimens eroded-corroded at 2.5 m/s for 100 h by 90/10 and 70/30 $SiO_2/CaSO_4$ mixtures. It confirms the possibility that sulfidation of carbon steel could occur at a temperature of 450 °C when in the presence of a $CaSO_4$ deposit at a low particle impact velocity of 2.5 m/s in a considerably longer duration test. At higher particle velocity conditions, other sets of factors determined the composition and morphology of the scale layer and no sulfide compound was identified.

The initiation of sulfidation between the scale layer and the substrate requires sufficient reaction time and exposure to a low $p_{O_2}$ atmosphere in the presence of $CaSO_4$ (Ref 13-15). Therefore, it is very difficult for the sulfidation to occur at high particle velocities where considerably more oxygen is directed to the target surface and the test is for a much shorter duration. Because a certain thickness of deposit/scale layer is required to develop the low $p_{O_2}$ atmosphere at the metal substrate surface, iron sulfide did not form on the samples eroded by $SiO_2/CaSO_4$ ratios above the transition ratio because the high $SiO_2$ content caused sufficiently high erosion-corrosion rates to keep the scale layer too thin to have the low $p_{O_2}$ occur. This same logic generally applies to in-service exposures, especially in CFBCs, where the steady-state scale thickness is too thin to permit a low $p_{O_2}$ atmosphere to occur beneath it, but not necessarily always.

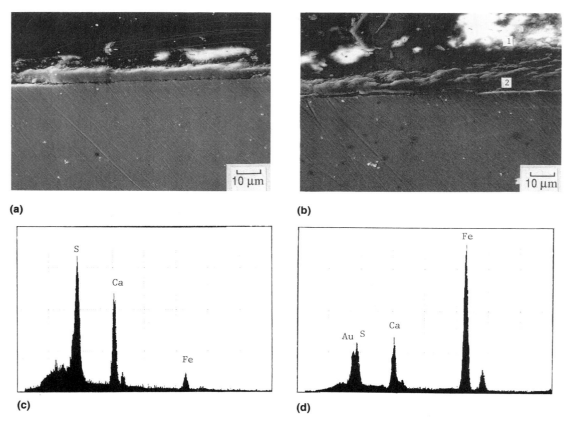

**Fig. 6.22** Cross sections and EDS peak analyses of 1018 steel tested with two $SiO_2/CaSO_4$ mixtures. $T = 450$ °C, $V = 2.5$ m/s, $t = 100$ h, $\alpha = 30°$. (a) $SiO_2/CaSO_4$: 90/10. (b) $SiO_2/CaSO_4$: 70/30. (c) Composition of point 1. (d) Composition of point 2

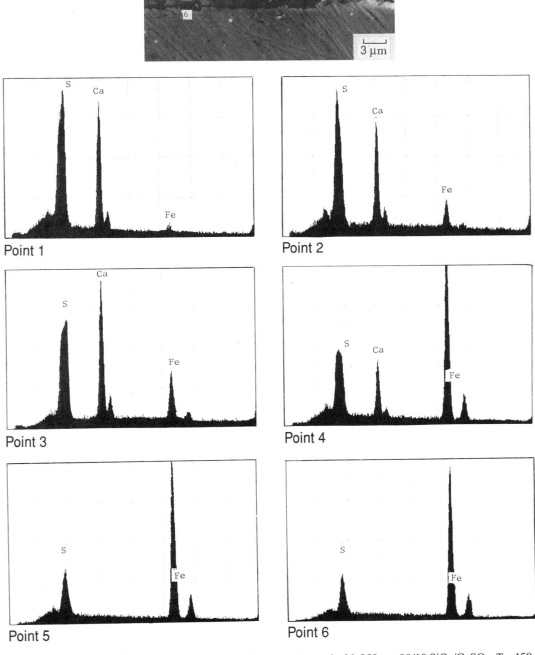

**Fig. 6.23** Cross section and EDS peak analyses of 1018 steel tested with 250 μm 90/10 SiO₂/CaSO₄. $T = 450$ °C, $V = 2.5$ m/s, $t = 100$ h, $\alpha = 30°$

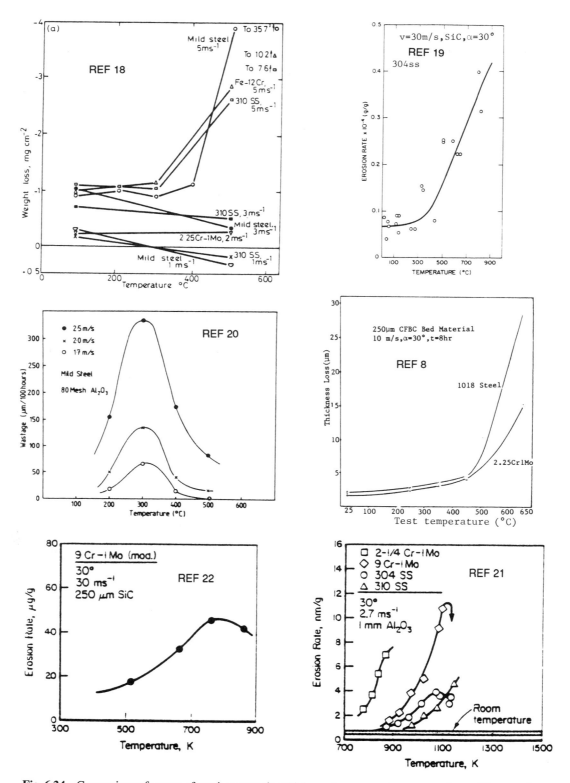

**Fig. 6.24** Comparison of curves of erosion-corrosion thickness loss versus temperature from different references

Some investigators (Ref 16) have divided elevated-temperature erosion-corrosion of metals into several somewhat arbitrary categories to be able to assign equations to each designated behavior that defines the rate of change in the oxide scale thickness and, hence, material loss. These divisions of the erosion-corrosion process are: erosion of oxide only, erosion-enhanced corrosion (three subtypes), oxidation-affected erosion and erosion of metal only. The dividing up of the erosion-corrosion process required that the oxide scale layer be the pure product of base metal oxidation and be either a constant thickness or in a spalling mode.

As can be seen in the micrographs in this section, the compositions of the scale layers are not simple substrate metal oxide scale, but rather an intimate, complex mixture of bed material and base metal oxide. The mixtures occur over the whole spectrum of alloys and exposure conditions experienced in carbon fuel combustion systems. This essentially negates the efforts to categorize the erosion-corrosion process into various arbitrary segments, such as those described in Ref 16.

## Erosion-Corrosion of Steel at Low Particle Velocities(Ref 17)

There have been several laboratory studies of erosion-corrosion of in-bed heat-exchanger tubing steel at the low particle velocities (for example, 2.5 m/s), that occur in the bed area of FBCs (Ref 18-24). Data from these studies is presented in Fig. 6.24. The studies primarily used hard erodents such as quartz and alumina. The group under Stott and Wood at the University of Manchester obtained curves of metal wastage versus

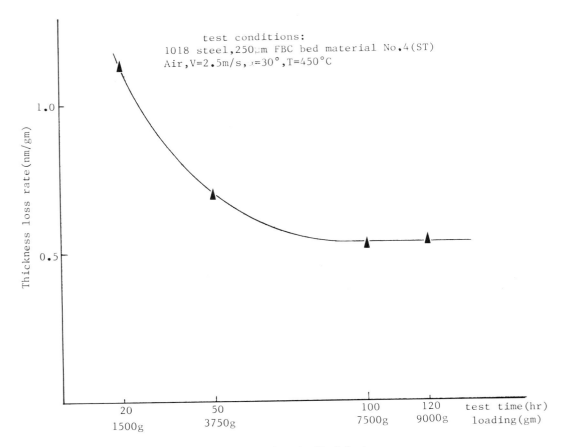

**Fig. 6.25** Incremental eroded-corroded rate of 1018 steel at $V = 2.5$ m/s

temperature for particle velocities of 1 to 5 m/s (Ref 18), which were the same shape as those of Levy's group at LBL which were generated at higher velocities (Ref 8, 19). In these studies there was a continuous increase in metal wastage with test temperature. Hutchings' group at the University of Cambridge tested the same steels in the same low particle velocity and elevated-temperature regime as Stott and got curves (Ref 20) that were markedly different, that is, the curves had a peak value. The shape of Hutchings' curves were more similar at higher temperatures to some work done by Sethi (Ref 21) and Levy's work (Ref 22) for a modified 9Cr-1Mo steel, using SiC as the erodent. They differed from Sethi's work at lower temperatures that are of prime importance in the operation of FBCs. Sethi's lower temperature behavior patterns matched Stott's and Levy's, an ever-increasing amount of loss with test temperature. The four groups used different test devices.

### Test Conditions

The tests were carried out in the elevated-temperature nozzle tester (Fig. 5.1, modified with a straight-drop tube replacing the mixing chamber). Air was utilized as the carrier gas for the particles, creating a generally oxidizing atmosphere. The target material for the erosion-corrosion tests was 1018, plain carbon steel, the same steel used for heat exchanger tubes in FBCs. Tests were carried out at six different temperatures: 25, 150, 250, 450, 500, and 550 °C at $V$ = 2.5 m/s using 250 μm diam bed material erodent particle No. 4 obtained from an operating CFBC. A test time of 100 h was used at a particle loading of 7500 g, which resulted in steady-state material loss conditions (Fig. 6.25). The test conditions nearly simulated those experienced by in-bed FBC heat exchanger tubes. In addition, to determine the effect of more erosive particles and a somewhat higher particle velocity on the temperature effect, 180 μm angular $Al_2O_3$ was also used to erode the 1018 steel in the temperature range from 25 up to 550 °C for an exposure period of 8 h at a 30° impact angle and $V$ = 10 m/s.

The specimen metal wastages were determined by the thickness change of cross sections taken through the central part of the erosion zone using microscopic micrometer measurements, described earlier in this chapter.

### Effect of Temperature

Figure 6.26 is the curve of test temperature versus thickness loss of 1018 steel eroded-corroded at a low particle velocity of 2.5 m/s at a 30° impact angle by 250 μm CFBC bed material for an exposure time of 100 h. It can be seen that there are four different regimes. The first regime occurred at temperatures to about 200 °C in which the erosion-corrosion metal wastage was very low and had essentially no dependence on temperature. In the second regime, the metal wastage sharply increased from about 200 °C to a maximum rate at 350 °C. In the third regime, above 350 °C, the wastage rate decreased with temperature to 500 °C. Above 500 °C the erosion-corrosion metal wastage dramatically increased with increasing temperature. The shape of the curve to 500 °C was the same as others have reported (Ref 20, 21, 23). However, the loss rate did not fall to as low a level as occurred in other investigations. The sharp increase in rate of metal loss above 500 °C is completely different from the other findings.

In the low-temperature regime, the very low wastage was due to mechanical erosion at a very low particle velocity. In the second regime, where the loss rate sharply increased, oxidation played a major role in the metal wastage with the rate of oxidation being greatly enhanced by the particle impacts, as described in Chapter 5. Increasing oxide plasticity and thickness, providing protection to the substrate metal, are responsible for the wastage decrease with rising temperature in the third regime. The sharp increase in metal wastage at temperatures >500 °C is due to spalling of the thick scale that built up on the 1018 steel at temperatures above its free oxidation temperature.

Figure 6.27 shows the thickness loss versus temperature for 1018 steel eroded by 180 μm angular $Al_2O_3$ at a higher particle velocity, $V$ = 10 m/s. It can be seen that at these test conditions, the erosion-corrosion metal wastage curve continually increased with increasing temperature without showing the peak that was observed by

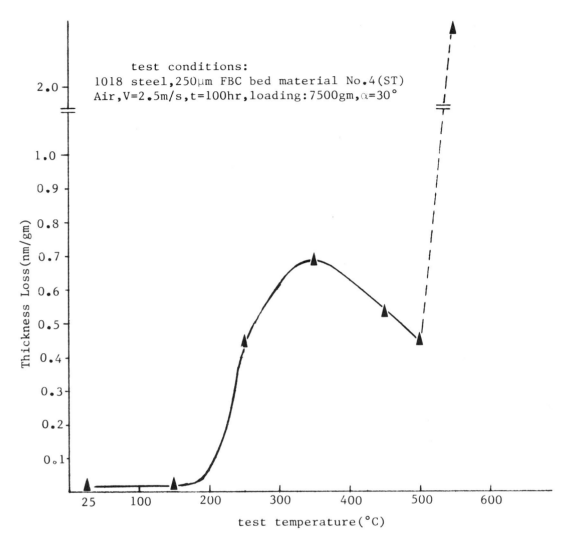

**Fig. 6.26** Test temperature versus thickness loss for 1018 steel eroded-corroded at $V = 2.5$ m/s

Hutchings and Sethi when they used $Al_2O_3$ erodent at $V = 2.5$ m/s (Ref 20, 21) (see Fig. 6.24).

When bed material was used to erode 1018 steel (see Ref 8 curve in Fig. 6.24), at low elevated temperatures from 200 to 450 °C, a relatively low rate of increase in metal wastage occurred. Increasing the temperature from 450 to 650 °C caused a large increase in metal wastage. At temperatures below 450 °C, corrosion (oxidation) per se, essentially did not occur in the metal wastage process of 1018 steel, whereas it had a significant effect above 450 °C.

Figure 6.27 shows that when $Al_2O_3$ was used as the erodent rather than bed material, there was also little change in the erosion-corrosion metal wastage rate at low temperatures (below about 200 °C). In this regime, corrosion did not occur and metal loss was by straight mechanical erosion. When temperatures above 250 °C were used with the $Al_2O_3$ erodent, the erosion-corrosion metal wastage greatly increased because oxidation played a major role. The key difference was the increase in the temperature at which increased loss rates occurred from 200 °C when $Al_2O_3$ particles were used to 450 °C when bed material particles were used (Ref 8). The bed material particles tended to deposit and adhere to the surface of the metal, promoting the formation of a protective layer against oxidation and metal wastage while the $Al_2O_3$ particles did not. This increased the temperature at which erosion-corrosion metal wastage began to increase with increasing temperature.

It can be seen in Figure 6.26 and 6.27 that particle velocity influenced the relationship between metal wastage and test temperature. At the low particle velocity of 2.5 m/s, a peak in the metal wastage curve occurred at approximately 350 °C. At a particle velocity of 10 m/s, no peak in the metal wastage curve occurred. This coincides with the data reported in Ref 18, in which

Stott found that at 5 m/s the damage sustained by mild steel and 2.25Cr-1Mo steel in a fluidized bed of coarse silica particles increased with increasing temperature. However, using fine particles or lower velocities caused the wastage to decrease with increasing temperature to 500 °C (see Fig. 6.24).

At lower particle velocities in the third regime (see Fig. 6.26), the oxide behaved in a more ductile manner with increasing temperature, resulting in lower metal wastage. At higher particle velocities, the oxide did not become ductile enough to reduce erosion damage as the temperature was increased. An explanation of this difference in the nature of the scale may be found in

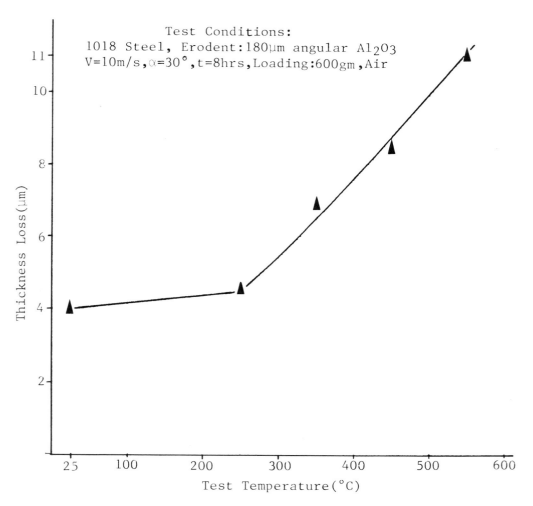

**Fig. 6.27**   Test temperature versus thickness loss for 1018 steel eroded-corroded at $V = 10$ m/s

the pseudoplasticity of oxide scale at elevated temperatures (Ref 24). The plasticity of oxide scale at certain elevated temperatures is related to the strain rate. At low strain rates, corresponding to $V = 2.5$ m/s, the microcracking of the oxide scale essentially takes place simultaneously with crack healing to cause macroscopic plasticity of the oxide, that is, pseudoplasticity occurred. Because plastic deformation can occur before cracking, the oxide scale can have a protective effect.

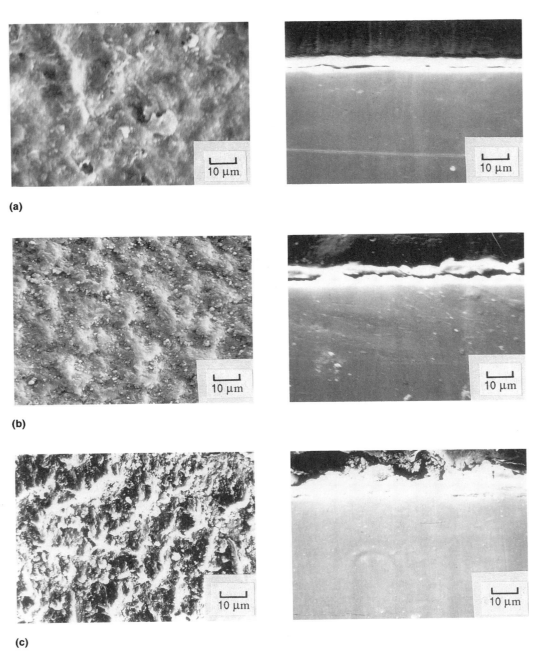

**Fig. 6.28**  Surface and cross section of 1018 steel specimens eroded-corroded by 250 μm CFBC bed material particles. $V = 2.5$ m/s, $t = 100$ h (7500 g), $\alpha = 30°$. (a) 250 °C. (b) 350 °C. (c) 450 °C

However at higher strain rates, corresponding to $V$ = 10 m/s, the crack healing could not take place, microcracks developed in the oxide and eventually cracking, chipping, or spalling occurred, as described in Chapter 2. The above explanation for the difference in behavior correlates well with the observed morphologies of the affected surfaces and cross sections of the eroded-corroded specimens.

### Microstructural Analysis

Figure 6.28 shows surfaces and cross sections of specimens eroded-corroded by CFBC bed ma-

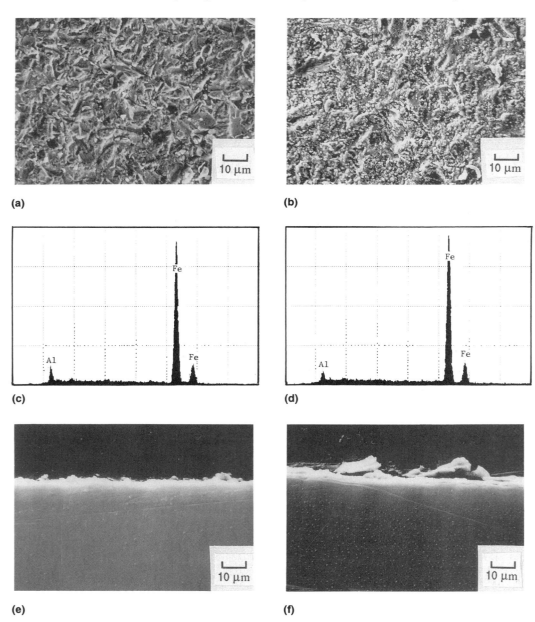

**Fig. 6.29** Surface and cross section of 1018 steel specimens eroded-corroded by 180 μm angular Al₂O₃. $V$ = 10 m/s, $t$ = 8 h (600 g), $\alpha$ = 30°. (a) Surface ($T$ = 250 °C). (b) Surface ($T$ = 450 °C). (c) Composition of surface ($T$ = 250 °C). (d) Composition of surface ($T$ = 450 °C). (e) Cross section ($T$ = 250 °C). (f) Cross section ($T$ = 450 °C)

terial at $V = 2.5$ m/s, at an impact angle of 30°, for 100 h, at 250, 350, and 450 °C. It can be seen that as the temperature was increased, the scale layer thickness increased. This indicated that the oxidation rate had increased. At 250 and 350 °C, the oxide scale did not become ductile enough to reduce erosion damage, and cracks formed and developed. At 350 °C, where the peak metal wastage rate occurred, cracks that formed in the oxide have extended down to the scale/metal interface. However, at 450 °C where metal loss was reduced, the oxide became more ductile, and the oxide scale remained continuous without fracture, as the cross section shows.

The morphologies and EDS analyses of specimens eroded-corroded at the higher particle velocity of 10 m/s, by angular $Al_2O_3$ particle are shown in Fig. 6.29. It can be seen that with increasing temperature the oxide scale thickness increased, but even at 450 °C the scale did not have enough ductility to prevent cracks from forming, causing a continual increase in metal wastage. Comparing Fig. 6.29 with Fig. 6.28, it can be seen that the scale formed at the lower particle velocity was thicker, denser, and more continuous than that formed at the higher particle velocity. Moreover, at 450 °C the scale formed at the low particle velocity became more ductile, resulting in more deformation and retention, while the scale formed at the higher particle velocity was still brittle, cracked, and was more readily removed.

### Effect of Impact Angle

Figure 6.30 shows the curve of the impact angle versus thickness loss of 1018 steel eroded-corroded at $V = 2.5$ m/s, $T = 450$ °C by 250 μm FBC bed material. The metal wastage at shallow angles was higher than that at steep angles. The highest wastage occurred at an impact angle of 45°. The general shape of the erosion curve is similar to that of a ductile metal except that the angle of highest metal loss was considerably higher than normally occurs in a ductile metal. This behavior is further evidence of the ductile nature of the oxide scale at 450 °C at the low impact velocity.

## Relationship between Erosion-Corrosion Metal Wastage and Temperature of In-Bed Heat Exchanger Tubes[Ref 25]

The surface temperature of in-bed heat exchanger tubes in bubbling fluidized bed combustors (BFBCs) has a great influence on the erosion-corrosion metal wastage (Ref 26-28). The relationship between erosion-corrosion metal wastage and surface temperature is complex and very dependent on the service or test conditions, the rate of oxidation of the specimen, and the properties of the oxide surface layer that can form, such as its composition, deformability, adhesion, and cohesion (Ref 29). In commercial BFBCs it has been consistently observed that at low metal temperatures (≤250 °C), wastage rate is relatively high and there appears to be little dependence of wastage on temperature. Somewhere above this temperature, typically close to 400 °C, the wastage rate decreases rapidly as the metal temperature increases over a narrow temperature range, which can be regarded as a transition temperature (Ref 28). In Ref 20, which reported on heat exchanger tube behavior in an operating BFBC, it was stated that "it is believed that this sharp drop is related to the formation of a protective oxide layer on the surface, but this had not been demonstrated."

Previous to the work reported herein, this behavior had never been duplicated in laboratory tests where the more controlled conditions could make it possible to gain an understanding of why the behavior pattern occurred. In the current work, a series of erosion-corrosion tests was carried out on 1018, 2.5Cr-1Mo, and an experimental higher silicon steel (2.5Cr-0.55Mo-1.4Si) (Table 6.6) at $V = 2.5$ m/s using 180 μm angular $Al_2O_3$ to investigate the relationship between erosion-corrosion metal wastage and temperature and to successfully obtain the pattern of metal wastage behavior that has been observed in operating BFBCs. The test conditions were the same as those used to determine the effect of low particle velocities described earlier in this chapter.

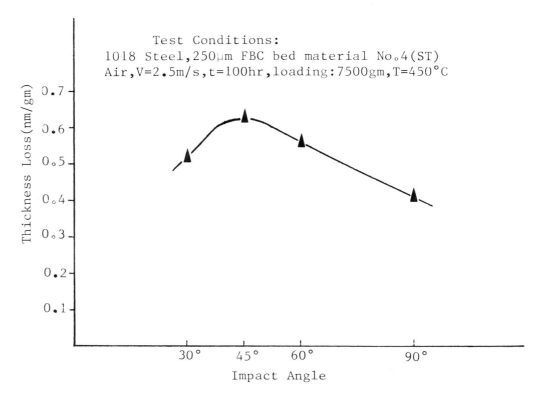

**Fig. 6.30** Impact angle versus thickness loss of 1018 steel specimens

**Table 6.6** Composition of three steels used in low-velocity erosion-corrosion tests

| Steel | Typical steel composition, wt% | | | | | | | |
|---|---|---|---|---|---|---|---|---|
| | Fe | C | Cr | Mo | Mn | Si | S | P |
| 1018 | bal | 0.14-0.20 | ... | ... | 0.6-0.9 | ≤0.60 | ≤0.05 | ≤0.04 |
| 2.25Cr-1Mo | bal | 0.15 | 2.0-2.5 | 0.9-1.1 | 0.3-0.6 | ≤0.5 | ≤0.03 | ≤0.03 |
| 2.5Cr-0.55Mo-1.4Si | bal | 0.1 | 2.5 | 0.55 | 0.45 | 1.4 | ≤0.02 | ≤0.03 |

### Thickness Loss-Temperature Curve

It can be seen in Fig. 6.31 that there are different relationships between erosion-corrosion metal wastage and temperature for 1018 and 2.25Cr-1Mo steels compared to 2.5Cr-0.55Mo-1.4Si steel. The metal wastage of the 1018 and 2.25Cr-1Mo steels increased with increasing temperature, reaching a maximum at 350 °C and then decreased with further temperature increases, as shown earlier in Fig. 6.26. Above 500 °C, the erosion-corrosion metal wastage dramatically increased with increasing temperature. However, for the 2.5Cr-0.55Mo-1.4Si steel, below 450 °C, the metal wastage was relatively high and constant, having little dependence on the temperature. Above 450 °C, the metal wastage decreased with test temperature as has been consistently observed on in-bed tubes in BFBCs (Ref 26, 28).

### 1018 and 2.25Cr-1Mo Steels

For 1018 and 2.25Cr-1Mo steels, the shape of the temperature-metal wastage curves indicates four different regimes had occurred that were described in Fig. 6.26. Comparing the curve of 1018 steel in Fig. 6.31 to that in Fig. 6.27, it can be seen that when tested at $V$ = 2.5 m/s and 550 °C the metal wastage had dramatically increased to 19.9 μm, while when tested at a higher velocity, $V$ = 10 m/s, at the same temperature, it had

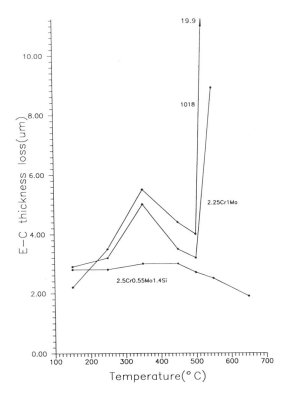

**Fig. 6.31** Temperature versus thickness loss of three steels eroded-corroded at $V = 2.5$ m/s

that in the temperature regime below the transition temperature (Ref 28), there is almost no scale on the surface, indicating that the oxide layer formed slowly and was not protective. This permitted the base metal to be directly eroded at a relatively high, more or less constant rate (Ref 19). However, in the higher-temperature regime, corresponding to the marked drop in thickness loss, the surfaces were covered with a thick, protective oxide layer. As the cross sections show, the oxide scale was continuous and dense without evidence of fracture, which indicated that during the tests the oxide scale formed in this temperature regime was more protective and ductile, that is, pseudoplasticity occurred. This resulted in more deformation and retention; thereby protecting the surface.

This occurrence is also indicated by the surface morphology of the 2.5Cr-0.55Mo-1.4Si steel specimens tested, which are shown in Fig. 6.33. When combined with the 450 °C cross section in Fig. 6.32, it can be seen that when tested at 450 °C the very thin scale on the surface of the specimen exhibited a brittle behavior with cracking and chipping, as evidenced by a fine surface texture consisting of many small craters and areas where bare metal was exposed. However, when tested at 650 °C a very coarse surface texture occurred with large craters and indentations occurring in the continuous surface scale. This indicates the ductile, protective nature of the thick scale that was present on the specimen surface.

The higher oxidation resistance of 2.5Cr-0.55Mo-1.4Si steel compared to 2.25Cr-1Mo steel (Ref 30) retarded scale formation at lower temperatures resulting in the exposure of base metal that eroded at a relatively high rate. When the temperature was high enough for oxidation to occur, the oxide that formed had sufficient ductility to undergo plastic deformation before cracking, and the wastage fell to a low level. Therefore, there was no increase in the slope of the curve for the 2.5Cr-0.55Mo-1.4Si steel, as occurred in the second regime for 1018 and 2.25Cr-1Mo steels (see Fig. 6.31). Also, the temperature range used in the present work is lower than the free oxidation temperature of 2.5Cr-0.55Mo-1.4Si steel (Ref 30). Therefore, no spall-

only increased to 11 μm. This difference is primarily due to the difference in test times. The specimen tested at 2.5 m/s was exposed for 100 h, while at 10 m/s it was only exposed for 8 h. At 550 °C, which is above the free-oxidation temperature of 1018 steel, the metal wastage was due to spalling of the thick scale. In this case, corrosion (oxidation) played a dominant role, and the thick, brittle oxide scale broke off in relatively large pieces. Because this mechanism occurred over a much longer period in the 100 h test than in the 8 h test, the metal wastage at 550 °C, was greater for the longer test at $V = 2.5$ m/s.

### 2.5Cr-0.55Mo-1.4Si Steel

Figure 6.32 is an expanded curve of thickness loss versus test temperature of 2.5Cr-0.55Mo-1.4Si steel specimens along with cross-section micrographs corresponding to the different test temperatures and thickness losses. It is evident

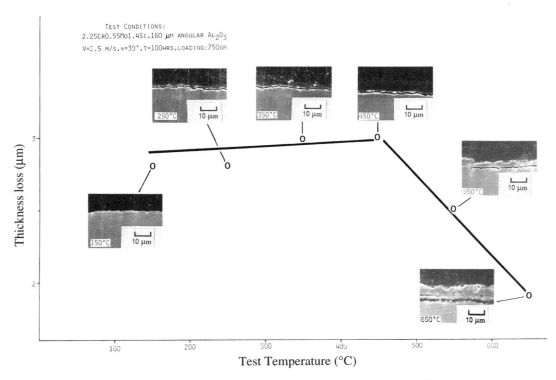

**Fig. 6.32** Cross sections of 2.5Cr-0.55Mo-1.4Si steel specimens corresponding to different test temperatures and thickness losses

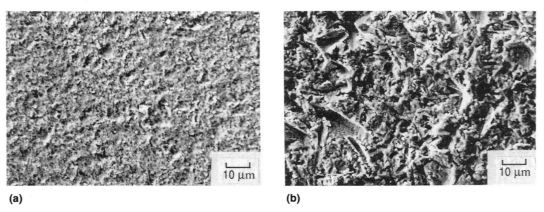

**(a)**                    **(b)**

**Fig. 6.33** Surfaces of 2.5Cr-0.55Mo-1.4Si steel specimens eroded-corroded by 180 μm angular Al₂O₃ particles. $V = 2.5$ m/s, $t = 100$ h (7500 g), $\alpha = 30°$. (a) $T = 450$ °C. (b) $T = 650$ °C

ing of thick scale occurred, and there was no resulting dramatic increase in metal wastage, as occurred for the two normal silicon content steels.

The difference in erosion-corrosion behavior between the 2.25Cr-1Mo and the 2.5Cr-0.55Mo-

1.4Si steels, alloys whose only essential difference is 1% more Si in the latter steel, provides important information to understanding the nature and role of the protective layer that can form on steels during erosion-corrosion and markedly affect their material wastage rates. It is known (Ref 30) that the addition of small amounts of additional silicon to chromium steels increases their resistance to oxidation. The increased resistance to the formation of an oxide layer of the 2.5Cr-0.55Mo-1.4Si steel prevented the formation of a protective scale at temperatures below the transition temperature of 450 °C and, therefore, a higher rate of material wastage of a bare metal surface occurred than occurred above the transition temperature (see Fig. 6.31 and Fig. 6.32). Below the 450 °C temperature, the normal silicon content, 2.25Cr-1Mo steel had a considerably different pattern of material loss that was related to the nature of the oxide scale that formed on its surface.

Above the 450 °C transition temperature, where the normal silicon content steel had dropped to a lower loss level and subsequently almost catastrophically increased its material wastage, the additional silicon steel was forming a ductile, tenacious oxide scale that continued to increase its protective capability to the maximum test temperature of 650 °C. This behavior difference can be related to the morphological differences in the scales. The exact differences in the protective nature of the scales on the essentially same chromium content steels are not known. But the significant differences in erosion-corrosion behavior of the two, nearly identical steels is indicative of the major role that the protective layer can play and the sensitive nature of its formation.

## Erosion-Corrosion of Cooled Steel(Ref 31)

Several studies have been carried out in an attempt to understand the erosion-corrosion phenomenon at elevated temperatures under conditions simulating those in FBCs (Ref 32-34). The tests were primarily conducted in laboratory apparatus, using isothermal specimens. However, in-service heat exchanger tubes in both FBCs and pulverized coal fired boilers are cooled by the water and/or steam being heated. The tubing is subjected not only to continuous surface particle impacts and the temperature of a hot gas-particle mixture, but also has both a temperature gradient across the thickness of the tube wall, and a posi-

tive heat flux during operation. Therefore, the erosion-corrosion material wastage and mechanism in actual heat exchanger tubing could be quite different from those determined in isothermal laboratory experiments. A series of erosion-corrosion tests was carried out on cooled 1018 steel specimens at elevated temperatures in order to investigate the effects of the temperature differences between the gas-particle mixture and the surface of specimens on their erosion-corrosion wastage rates and mechanisms.

### Test Conditions

The tests were carried out in an elevated-temperature, nozzle tester (see Fig. 5.1). Air was utilized as the carrier gas for the particles, creating a generally oxidizing atmosphere. The target material for the erosion-corrosion tests was 1018, plain carbon steel in the cold-rolled condition.

The specimen was mounted on a special specimen holder, which can cool one side using flowing water, while the opposite side is directly exposed to the environmental temperature and erodent particle flow. The environmental temperature ($T_1$) is defined as the temperature of the gas-particle mixture impacting the cooled specimen surface. Figure 6.34 is a schematic of the cooled specimen holder design with a cooled specimen and two thermocouples in place. Water passes through a tube at various flow rates to control the cooling efficiency and keep the specimen at the desired temperature. The water passes along the interior surface of the specimen, past one baffle to create turbulent flow, and then exits the holder through another tube.

The actual temperature of the specimen, $T_2$, was measured with a thermocouple inserted into a blind hole drilled in the back of the specimen. This arrangement was found to be sufficiently accurate for thin metal specimens. Another thermocouple was placed at the exposed surface, near the outer edge of the specimen, in a shelter that protected it against particle impacts, to measure the environmental temperature, $T_1$.

Tests were carried out at environmental temperatures ($T_1$) of 450 and 550 °C, at a particle velocity of 20 m/s, at impact angles of 30° and 90°, using 250 μm silica sand particles. The

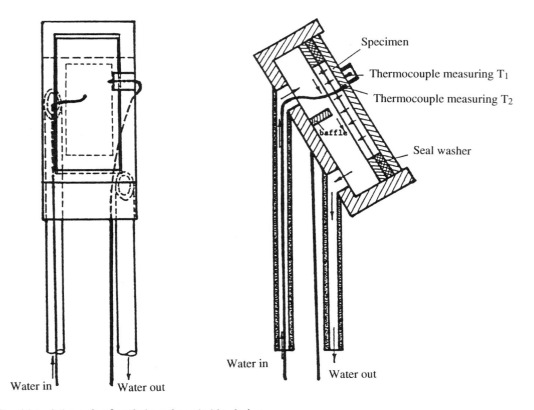

**Fig. 6.34** Schematic of cooled specimen holder design

specimens were cooled with water to result in various effective metal surface temperatures ($T_2$) from room to the environmental temperatures, $T_1$. A test time of 5 h was used at a particle loading of 375 g, which resulted in steady-state material loss conditions (see Fig. 6.1b).

The specimen metal wastages were determined by measuring the weight losses and thickness losses. The thickness losses were measured using two methods. One used a profilometer to measure the thickness of specimens before and after testing. The other measured the thickness of the sound material after testing using an optical micrometer accurate to 0.1 µm to observe a cross section through the central part of the erosion zone, as described at the beginning of this chapter.

### Erosion-Corrosion Metal Wastage

Table 6.7 lists the test results of the cooled 1018 steel specimens. The general behavior of the

steel can be seen in Fig. 6.35. However, a particularly important relationship that is indicative of the effect of the water cooling is best seen in Table 6.7. Comparing the two data points for $T_2$ = 25 °C for both the 30° and 90° impact angles show that the 25 °C surface temperature in the cooled specimen ($T_1$ = 450 °C) lost material at twice the rate of the uncooled 25 °C specimen tested at $T_1$ = 25 °C. The greater material wastage of the cooled specimen was also reported in Ref 35, but in that work the difference was reported to be several orders of magnitude.

Figure 6.35 is the plot of the specimen surface temperatures ($T_2$) versus thickness losses of the cooled 1018 steel specimens tested at environmental temperatures ($T_1$) of 450 and 550 °C. As the surface temperature of the specimen increased, the thickness loss reached a peak below $T_2$ = 100 °C and then decreased. This behavior is opposite to that which occurs in isothermal erosion-corrosion tests in this particle velocity regime where there is a continuous increase in

**Table 6.7    The erosion-corrosion wastage of cooled 1018 specimens**

| Temperature, °C | | Impact angle, $\alpha$ | Weight loss, mg | Thickness loss (profile), μm | Thickness loss (micro), μm | Difference, μm |
|---|---|---|---|---|---|---|
| environment $T_1$ | sample $T_2$ | | | | | |
| 25 | 25 | 30° | 0.4 | 4.27 | 4.4 | 0.13 |
| 450 | 450 | 30° | 10.8(a) | 5.59 | 7.4 | 1.81 |
| 450 | 300 | 30° | −4.4(a) | 6.15 | 8.2 | 2.05 |
| 450 | 200 | 30° | 0.3 | 7.85 | 9.6 | 1.75 |
| 450 | 150 | 30° | 4.2 | 9.52 | 11.1 | 1.58 |
| 450 | 100 | 30° | 4.9 | 10.32 | 11.7 | 1.38 |
| 450 | 50 | 30° | 7.3 | 14.88 | 15.3 | 0.42 |
| 450 | 25 | 30° | 3.7 | 8.68 | 9.0 | 0.32 |
| 25 | 25 | 90° | 0.1 | 3.18 | 3.3 | 0.12 |
| 450 | 450 | 90° | −2.5(a) | 3.38 | 4.4 | 1.02 |
| 450 | 350 | 90° | −3.0(a) | 3.78 | 5.0 | 1.22 |
| 450 | 300 | 90° | −1.6(a) | 4.94 | 5.9 | 0.96 |
| 450 | 250 | 90° | −1.7(a) | 5.46 | 6.5 | 1.04 |
| 450 | 200 | 90° | −1.5(a) | 6.76 | 7.1 | 0.34 |
| 450 | 150 | 90° | −1.5(a) | 7.04 | 7.4 | 0.36 |
| 450 | 90 | 90° | −0.8(a) | 7.14 | 7.6 | 0.46 |
| 450 | 25 | 90° | 1.8 | 6.35 | 6.7 | 0.35 |
| 550 | 550 | 30° | 24.0 | 37.31 | 40 | 2.69 |
| 550 | 450 | 30° | −6.2(a) | 5.7 | 7.8 | 2.10 |
| 550 | 350 | 30° | −5.8(a) | 6.08 | 8.1 | 2.02 |
| 550 | 250 | 30° | −3.9(a) | 7.78 | 9.5 | 1.72 |
| 550 | 150 | 30° | 0.9 | 12.88 | 13.4 | 0.52 |
| 550 | 75 | 30° | 13.0 | 15.08 | 15.8 | 0.72 |
| 550 | 35 | 30° | 4.3 | 10.51 | 10.9 | 0.39 |

Note: Test conditions: 250 μm sand, $V = 20$ m/s, loading 375 g, test time 5 h. (a) Weight gain

metal wastage with increasing temperature (see Fig. 6.27). A distinct break in the slope of the curves in Fig. 6.35 occurred near 250 °C with much less of a temperature effect on the metal loss occurring above 250 °C. The reasons for this behavior will be discussed later, after the micrographs of the test surfaces have been analyzed, as the surface morphology and composition relate to the reason for the behavior differences.

The solid lines in Fig. 6.35 show the metal wastage determined by measuring the cross-sectional thickness change using an optical micrometer. The dashed lines show the metal wastage determined by measuring the thickness change using a profilometer. It can be seen that the metal wastage measured by the former method is greater than by the latter method at both $T_1 = 450$ and 550 °C. The optical micrometer method is used as the primary method at LBL, while the profilometer method has been used by most other investigators (Ref 35). In this work, both methods were used so that the difference between them could be seen. The difference is primarily due to the fact that when measuring the cross-sectional thickness change, the thickness of oxide scale that forms must be considered. Meas-

uring the thickness change using a profilometer includes the thickness of the oxide in the measurement and, therefore, metal thickness loss data are less than the actual value. The optical micrometer accurately measures the amount of metal remaining in the cross section. Also, there is a basic measurement difference between the different measuring methods. Even though there was no scale on the surface of specimens tested at 25°/25 °C ($T_1/T_2$), a slight discrepancy between the two measuring methods (approximately 0.12 to 0.13 μm) still occurred (see Table 6.7).

From Fig. 6.35 it can be seen that when the specimen surface temperature ($T_2$) was increased, the metal wastage difference resulting from the two measurement methods increased. This means that with increasing surface temperature ($T_2$), the thickness of the scale on the surface of the specimens increased. The morphological evidence of this is shown in Fig. 6.36, which shows the cross sections of cooled 1018 specimens tested at $T_1 = 450$ °C and at different surface temperatures, $T_2$. It can be seen that when specimens were tested at higher $T_2$ (300 and 450 °C ), the scale was thicker and more continuous than those tested at lower $T_2$, even though it is still

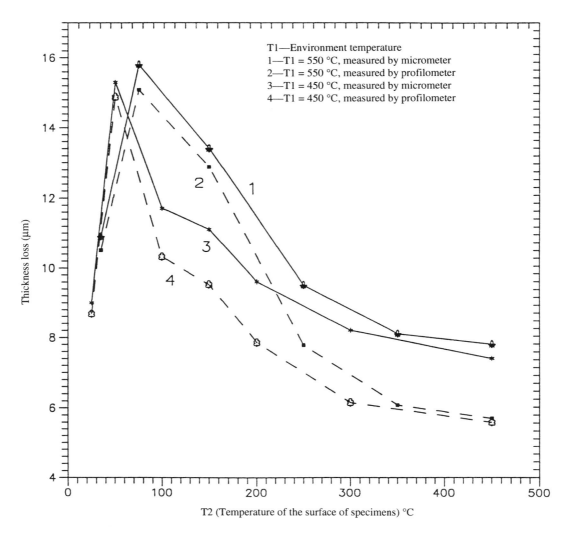

**Fig. 6.35** Temperature of $T_2$ versus thickness losses of cooled 1018 steel tested at $T_1$ = 450 and 550 °C

relatively thin. Some electron charging in the micrographs makes it somewhat difficult to see the scale in the photos.

From the data obtained by both measurement methods (see Table 6.7 and Fig. 6.35), it was found that the cooled 1018 steel specimens had higher metal wastages than the isothermal specimens at both impact angles of 30° and 90° when tested at $T_1$ = 450 °C. It also can be seen that the metal wastages were related to the temperature difference between the environment and the surface of the specimens. As mentioned above, with increasing temperature difference, the erosion-corrosion metal wastage increased, reaching a maximum at a specimen surface temperature of 50 to 90 °C, and then decreasing to 25 °C surface temperature. The $T_1$ = 550 °C tests had the same pattern except for the isothermal test. The isothermal specimen tested at 550 °C had a very high metal wastage (see Table 6.7). The reason for this is that the temperature of the isothermal specimen surface reached 550 °C, which is above the free oxidation temperature of 1018 steel, and the metal wastage dramatically increased due to spalling of the thick scale that formed (Ref 36).

Comparing curves 1 and 2 with curves 3 and 4 in Fig. 6.35, it can be seen that increasing the environmental temperature from 450 to 550 °C,

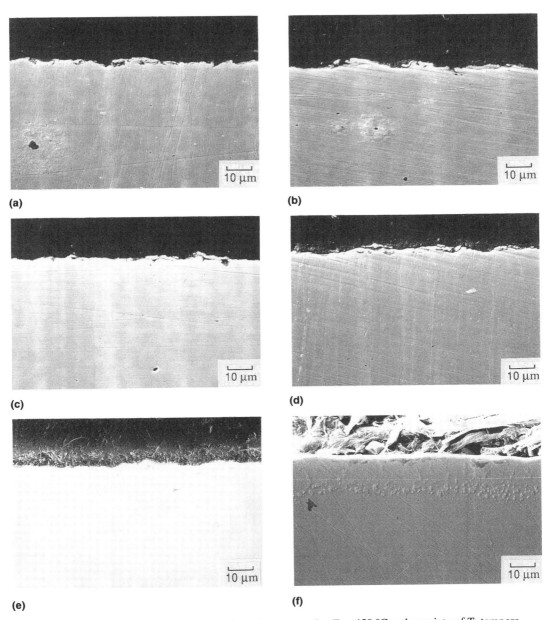

**Fig. 6.36** Cross sections of cooled 1018 steel specimens tested at $T_1 = 450$ °C and a variety of $T_2$ temperatures. 250 μm sand particles, $V = 20$ m/s, $t = 5$ h (375 g), $\alpha = 30°$. (a) $T_1/T_2 = 25/25$ °C. (b) 450/25 °C. (c) 450/50 °C. (d) 450/150 °C. (e) 450/300 C. (f) 450/450 °C. 250 μm sand particles, v = 20 m/s, t = 5 h (375 g), a = 30 °

increased the metal wastage of all specimens, even though their surface temperatures, $T_2$, were the same. The reason for this increase can be attributed to the fact that raising the environmental temperature increased the temperature difference between the environment and surface temperatures of the specimens, which has been established to cause greater metal wastage.

Figure 6.37 compares the metal wastages of specimens tested at an impact angle of 30° with those tested at an impact angle of 90°. The curves indicate that the cooled 1018 steel specimens tested at the conditions in this investigation had the erosion behavior of ductile materials. This correlates with the fact that, as Fig. 6.36 cross sections show, only a thin, dis-

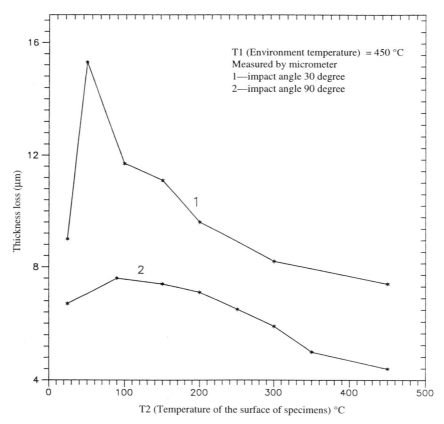

**Fig. 6.37** Comparison of the metal wastages of specimens test at $\alpha = 30°$ and $\alpha = 90°$

**Table 6.8** Chemical analysis of eroded-corroded surface of specimens by EDS

| Tested at $\alpha$ | $T_1/T_2$, °C/°C | Composition, wt% | | |
|---|---|---|---|---|
| | | Si | Al | Fe |
| 30° | 25°/25° | 4.96 | 0.18 | 94.85 |
| 30° | 450°/25° | 0.92 | 0 | 99.08 |
| 30° | 450°/50° | 1.13 | 0 | 98.87 |
| 30° | 450°/150° | 7.11 | 0 | 92.89 |
| 30° | 450°/300° | 22.62 | 5.37 | 72.01 |
| 30° | 450°/450° | 24.21 | 6.22 | 69.57 |
| 90° | 25°/25° | 11.36 | 1.82 | 86.82 |
| 90° | 450°/25° | 9.38 | 0.32 | 90.32 |
| 90° | 450°/90° | 16.19 | 2.57 | 81.24 |
| 90° | 450°/150° | 18.91 | 2.89 | 78.20 |
| 90° | 450°/250° | 39.25 | 7.20 | 53.55 |
| 90° | 450°/300° | 39.56 | 10.27 | 50.17 |
| 90° | 450°/350° | 42.74 | 10.78 | 46.48 |
| 90° | 450°/450° | 49.92 | 13.08 | 37.00 |
| 30° | 550°/35° | 1.41 | 0 | 98.59 |
| 30° | 550°/75° | 1.83 | 0 | 98.17 |
| 30° | 550°/150° | 10.58 | 1.22 | 88.20 |
| 30° | 550°/350° | 33.46 | 8.92 | 57.62 |

continuous oxide scale formed on the surface of the cooled 1018 steel specimens. This permitted the impacting particles to plastically deform the ductile base metal, directly involving it in the metal wastage behavior. Hence, it behaved as a ductile material, which generally undergoes higher erosion at a shallow impact angle than at a steep angle.

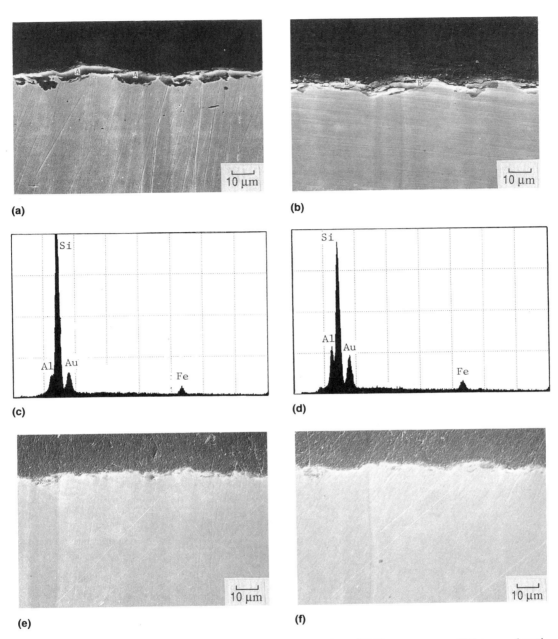

**Fig. 6.38** Cross sections of 1018 steel specimens tested at a number of $T_1/T_2$ temperatures. 250 µm sand particles, $V = 20$ m/s, $t = 5$ h (375 g), $\alpha = 90°$. (a) 450/350 °C. (b) 450/250 °C. (c) Composition of cross section (position A). (d) Composition of cross section (position B). (e) 450/150 °C. (f) 25/25 °C

Table 6.8 lists the surface compositions of the tested specimens. It can be seen that the amount of deposited erodent particles on the surface of specimens tested at a 90° impact angle is more than that on the specimen surfaces tested at a 30° impact angle. With increasing $T_2$ at the same $T_1$ or increasing $T_1$ at the same $T_2$, the amount of deposited $SiO_2$ from the sand erodent particles on the impacted surface of specimens increased at both impact angles tested. This indicated that at higher specimen surface temperatures ($T_2$) and environmental temperatures ($T_1$) the surface was more receptive to retaining erodent particles, probably due to a combination of increased oxide formation that trapped fine depositing particles and a general softening of the material surface. The deposit contributed to the reduction of metal wastage by protecting the surface.

### *Morphology*

Figure 6.38 and 6.39 show the cross sections and surfaces of specimens eroded-corroded at a 90° impact angle and $T_1 = 450$ °C at a variety of surface temperatures, $T_2$. Combined with Table 6.8, it can be seen that the surfaces of specimens tested at $T_2 = 250$ and 350 °C had a significant silicon content, while those tested at $T_2 = 25$ and 150 °C had little silicon on the surface. The white-appearing areas on the surfaces of specimens tested at 250 and 350 °C are thought to be areas of particularly high erodent particle content. This is based on the presence of high silicon and aluminum contents as determined by the EDS analyses listed in Table 6.8. They were not observed on the surfaces tested at lower $T_2$ temperatures. Note in Table 6.8 that there was a sharp increase in silicon content at a particular temperature. The appearances of micrographs of specimens tested at an environmental temperature, $T_1$, of 550 °C are very similar to those from the $T_1 = 450$ °C tests.

Figure 6.40 shows the surfaces of the specimens eroded-corroded at a 30° impact angle and $T_1 = 450$ °C at various surface temperatures, $T_2$. It can be seen by using Fig. 6.39 and Fig. 6.40, that the morphologies of the surfaces of the cooled specimens tested at 30° impact at the larger temperature gradients (lower $T_2$) are basically similar to the specimens tested at 25/25 °C. A typical ductile metal erosion mechanism occurred that consisted of platelets and craters, with indentations and some large gouges and striations. Some evidence of cracks and chips in a thin scale were observed on the eroded-corroded surfaces of the specimen tested at 450/25 °C (see Fig. 6.40e). This indicates that the erosion-corrosion mechanism of those cooled specimens is the same as that of erosion of ductile metals at room temperature, despite the different metal wastages. Their material loss was predominantly by the direct erosion of the bare metal by the platelet

mechanism of erosion described in Chapter 2, with no protection from any kind of layer on its surface. This can be seen in the cross-section morphology shown in Fig. 6.36. There were no concentrated silicon and aluminum deposits on the tested surface at any $T_2$ temperature in the 30° impact tests.

At higher surface temperatures, generally 250 °C, the combined morphologies of surfaces and cross sections indicated that the deposit/scale was becoming continuous and thick enough to begin to protect the metal from erosion-corrosion metal wastage. This can be seen in the data in Table 6.7.

From Fig. 6.38 and 6.39, it is clear that the general morphologies of the eroded surfaces and cross sections of specimens tested at 90° impact angle differ little from those specimens tested at 30°, which means the erosion-corrosion mechanism does not depend strongly on the impact angle for cooled 1018 steel specimens.

Comparing the data in Table 6.7 with that in Table 6.8, it is worthy to note that the relationship between the metal wastage and the temperature gradient $T_1$-$T_2$, is similar to that between the amount of erodent particle deposit and the temperature gradient. As the amount of particle deposit increased with a reduction in the temperature gradient, the resulting improved protection reduced the metal wastage.

### *Explanation*

The reduced metal wastage that occurred as the amount of cooling was decreased (higher surface temperature, $T_2$) compared to the loss in the isothermal tests can be attributed to several factors. At the lower-surface temperatures, below the break in the curves of Fig. 6.35, the increase in ductility of the base metal with temperature reduced the ability of the erodent particles to concentrate their force at the point of impact, which reduced their erosivity; hence the decrease in metal loss. It is similar to the behavior of 1018 steel tested at elevated temperatures in nitrogen where essentially no oxidation occurred (Ref 19) (see Chapter 5). Above the temperature where there is a break in the curve in Fig. 6.35, the effect of the beginning of metal oxidation that increased the metal loss rate with tempera-

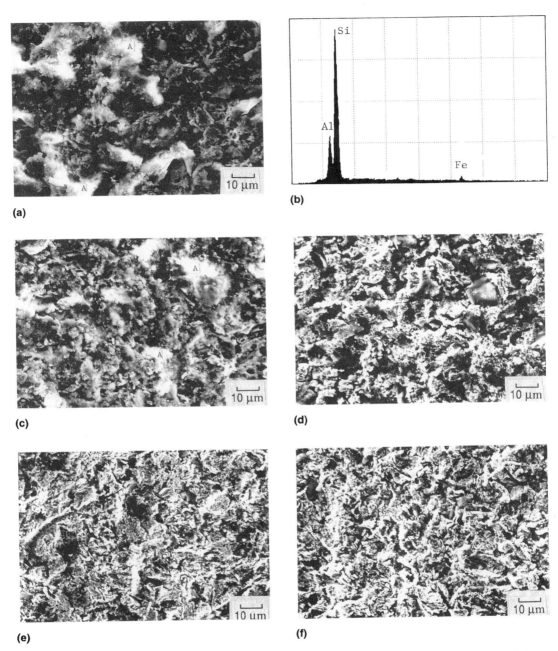

**Fig. 6.39** Surfaces of 1018 steel specimens tested at a number of $T_1/T_2$ temperatures. 250 μm sand particles, $V = 20$ m/s, $t = 5$ h (375 g), $\alpha = 90°$. (a) 450/350 °C. (b) Composition of surface (white area A). (c) 450/250 °C. (d) 450/150 °C. (e) 450/25 °C. (f) 25/25 °C

ture in earlier work (Ref 10) and the deposit-scale layer that decreased it combine to moderate the effect of ductility, resulting in a flattening of the curve.

The difference in the metal wastage rate at a surface temperature of $T_2 = 25$ °C in specimens tested at environmental temperatures of $T_1 = 25$ and 450 °C provides insight into the basic effect

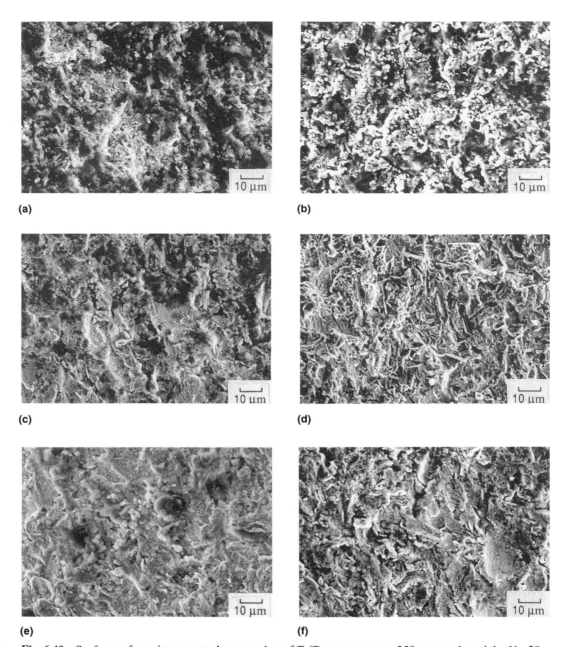

**Fig. 6.40** Surfaces of specimens tested at a number of $T_1/T_2$ temperatures. 250 μm sand particle, $V = 20$ m/s, $t = 5$ h (375 g), $\alpha = 30°$. (a) 450/450 °C. (b) 450/300 °C. (c) 450/150 °C. (d) 450/50 °C. (e) 450/25 °C. (f) 25/25 °C

of water cooling on the changes in metal wastage between isothermal testing and cooled specimen testing. It is thought that the severe plastic deformation that occurs at the point of impact of erodent particles causes significant heating of the surface, to near the annealing temperature (Ref

37), as discussed in Chapter 2. The temperature rise is thought, in part, to be related to the heat conduction through the metal, away from the impacted surface (Ref 38). Also, erosion of metals has been related to the ductility of the alloy being eroded (Ref 39). The ductility of a metal is related to its temperature with greater ductility generally occurring at higher temperatures (Ref 40).

Combining these behavioral aspects of metals, the following scenario is thought to have occurred and been responsible for the behavior of the cooled specimens compared to the isothermal specimens: The increase in the immediate surface temperature of the specimens that occurs as the result of localized plastic deformation upon impact was less in the cooled specimens than in the isothermally tested ones because part of the heat was drawn away by the cooling water through conduction. The lower immediate surface temperature resulted in the surface having less ductility and, therefore, a greater erosion rate occurred. The uncooled, isothermal specimen tested at 25 °C had a minimum amount of its heat from surface plastic deformation drawn away, and it retained the maximum surface ductility to resist metal wastage. This resulted in the $T_2 = 25$ °C isothermal test specimen having half the metal wastage of the $T_2 = 25$ °C cooled specimen in the $T_1 = 450$ °C tests.

As the cooling was reduced in the tests where the surface temperature, $T_2$, was nearer to the environmental temperature, $T_1$, more plastic deformation heating was retained on the surface, which resulted in increased ductility and lower metal wastage. Above $T_2 = 250$ °C, where there is sufficient surface oxidation to cause a deposit/scale layer to begin to form (see Fig. 6.36), the protection provided by the layer, along with the increased ductility, contributed to the lower metal wastage. This is indicated by the break in the curves in Fig. 6.35. However, at lower surface temperatures, $T_2$, near the peak in Fig. 6.35, the effect of the cooling was to draw just enough heat away from the surface to have the 1018 steel at an immediate surface temperature where a high strain age embrittlement could occur (Ref 41). The resulting decreased ductility resulted in increased metal wastage.

## References

1. Stringer, J.; Stallings, J.F.; and Wheeldon, J.M., Wastage in Bubbling Fluidized-Bed Combustors: an Update, *Proc. ASME 1989 Int. Conf. Fluidized Bed Combustion* (San Francisco), May 1989, p 857-862

2. Slusser, J.W.; Bixler, A.D.; and Bartlett, S.P., "Materials Experience From a Circulating Fluidized Bed Coal Combustor," Paper 90285, Corrosion 90 (Las Vegas), NACE, April 1990

3. Levy, A.V.; Wang, B.Q.; Geng, G.Q.; and Mack, W., "Erosivity of Particles in Circulating Fluidized Bed Combustors," Paper 543, Corrosion 89, NACE

4. Kalmanovitch, D.P.; Hajicek, D.R.; and Man, D.M., "Corrosion and Erosion in a Fluidized Bed Combustor: Effect of Bed Chemistry," Paper 547, Corrosion 89, NACE

5. Wang, B.Q.; Geng, G.Q.; and Levy, A.V., Surface Behavior of Heat Exchanger Tubes in Fluidized Bed Combustors, *Surf. Coat. Technol.*, Vol 42, 1990, p 253-274

6. Wang, B.Q.; Geng, G.Q.; Levy, A., "Fluidized Bed Combustor Particle Effects on the Erosion-Corrosion of Steel," Paper 34, EPRI Workshop on Materials Issues in Circulating Fluidized Bed Combustors, Electric Power Research Institute, June 1989

7. Kleist, D.M., "One Dimensional Two Phase Particulate Flow," Report LBL-6967, M.S. thesis, Lawrence Berkeley Laboratory, University of California, Berkeley, CA, 1977

8. Wang, B.Q.; Geng, G.Q.; and Levy, A.V., "Erosion-Corrosion of Tubing Steels at Simulated Fluidized Bed Combustor Convection Pass Conditions," Paper 15, Proc. NACE 4th Berkeley Conference on Corrosion-Erosion-Wear of Materials at Elevated Temperatures (Berkeley, CA), Feb 1990

9. Levy, A.V., and Man, Y.F., "Effect of Test Variables on the Erosion-Corrosion of Chromium Steel," Paper 17, Corrosion 87 (San Francisco), NACE, March 1987

10. Geng, G.Q., Wang, B.Q.; Levy, A.V., The Effect of Composition of Fluidized Bed Materials Combustor Beds on the Erosion-Corrosion of Carbon Steel, *Wear*, Vol 150 (No. 1, 2), Oct 1991, p 125-134

11. Geng, G.Q.; Wang, B.Q.; Levy, A.V., Effect of Calcium Sulfate Particles on Corrosion and Erosion-Corrosion of Carbon Steel, *Wear*, Vol 155, 1992, p 149-161

12. Levy, A.V.; Wang, B.Q.; and Geng, G.Q., Relationship between Feedstock Characteristics and Erosivity of FBC Bed Materials, *J. Eng. Gas Turbines Power (Trans. ASME)*, Vol 114, Jan 1992, p 145-151

13. Natesan, K., Corrosion of Heat Exchanger Materials in Simulated FBC Environments, *Corrosion*, Vol 38 (No. 7), July 1982, p 361-373

14. Natesan, K., "Corrosion of Alloys in FBC Systems," ORNL/FMP-86-1, Fossil Energy Material Program Processes Report, Oak Ridge National Laboratory

15. Natesan, K., "Corrosion of Alloys in Mix-Gas and Combustion Environments," ORNL/FMP-87/4, Fossil Energy Materials Program Conf. Proc., Oak Ridge National Laboratory

16. Rishel, D.M.; Pettit, F.S.; and Birks, N., "Some Principal Mechanisms in the Simultaneous Erosion and Corrosion Attack of Metals at High Temperature," Paper 16, Proc. NACE Conf. Corrosion-Erosion-

Wear of Materials at Elevated Temperatures (Berkeley, CA), 1990

17. Geng, G.Q.; Wang, B.Q.; Levy, A.V., Erosion-Corrosion of 1018 Steel Eroded at Low Velocities, *Wear*, Vol 155, 1992, p 137-147

18. Stott, F.H.; Green, S.W.; and Wood, G.C., "The Erosion-Corrosion Behavior of Low Alloy Steels Under Fluidized-Bed Conditions at Elevated Temperatures," Paper 545, Corrosion 89 (New Orleans), NACE, April 1989

19. Levy, A.V.; Yan, J.; and Patterson, J., Elevated Temperature Erosion of Steels, *Wear*, Vol 108 (No. 1), 1986, p 43-60

20. Ninham, A.J.; Hutchings, I.M.; and Little, J.A., "Erosion-Corrosion of Austenitic and Ferritic Alloys in a Fluidized Bed Environment," Paper 544, Corrosion 89, (New Orleans), NACE, April 1989

21. Cory, R.G., and Sethi, V.K., "Erosion-Corrosion of Engineering Alloys in Oxidizing Environment," Paper 10, Corrosion 87 (San Francisco), NACE, March 1987

22. Levy, A.V., Contributions to U.S. Department of Energy AR&TD Fossil Energy Materials Program, in Semi-Annual Reports from the Oak Ridge National Laboratory

23. Sethi, V.K., and Wright, I.G., Observations on the Erosion-Oxidation Behavior of Alloys, *Corrosion and Particle Erosion at High Temperatures*, Proc. Symp. (Las Vegas), TMS-ASM, March 1989, p 245-263

24. Schutze, M., Plasticity of Protective Oxide Scale, *Mater. Sci. Technol.*, Vol 6 (No. 1), Jan 1990, p 32-38

25. Wang, B.Q.; Geng, G.Q.; and Levy, A.V., The Relationship between Erosion-Corrosion Metal Wastage and Temperature, *Wear*, Vol 159 (No. 2), 1992, p 233-239

26. Tossaint, H.; Rademakers, P.; and Van Norden, P., Conf. paper, 7th Int. Conf. and Exhibition on Coal Technology and Coal Trade (Amsterdam), Nov 1988

27. Stringer, J. and Stallings, J., *Proc. ASME 1991 FBC Conf.* (Montreal), 21-24 April 1991, p 589-608

28. Holtzer, G.J., and Rademakers, P., *Proc. ASME 1991 FBC Conf.* (Montreal), 21-24 April 1991, p 743-753

29. Stott, F.H.; Stack, M.M.; and Wood, G.G., "The Role of Oxides in the Erosion-Corrosion of Alloys Under Low Velocity Conditions," *Proc. 4th Berkeley Conf. Corrosion-Erosion-Wear of Materials at Elevated Temperatures* (Houston), A.V. Levy, Ed., NACE, 1991, p 12-1 to 12-16

30. Geng, G.Q.; Wang, B.Q.; Hou, P.Y.; and Levy, A.V., "The Effect of Additional Silicon on the Corrosion and Erosion-Corrosion of Low Chromium Steels," *Wear*, Vol 150 (No. 1-2), 1991, p 89-105

31. Wang, B.Q.; Geng, G.Q.; and Levy, A.V., Erosion and Corrosion of Cooled 1018 Steel, *Wear*, Vol 161 (No. 1), 1993, p 41-52

32. Hutchings, I.; Little, J.; Ninham, A., "Low Velocity Erosion-Corrosion of Steels in a Fluidized Bed," *Proc. 4th Berkeley Conf. Corrosion-Erosion-Wear of Materials at Elevated Temperatures*, (Houston), A.V. Levy, Ed., NACE, 1991, p 14-1 to 14-17

33. Sethi, V.J., and Wright, J.G., *Corrosion & Particle Erosion at High Temperatures*, V. Srinivasan and K. Vedula, Ed., TMS, 1989, p 245-263

34. Wang, B.Q.; Geng, G.Q.; and Levy, A.V., *Proc. of Workshop on Materials Issues in Circulating Fluidized-Bed Combustors*, EPRI, 1990, p 34-1 to 34-38

35. Ninham, A.J.; Little, J.A.; Hutchings, I.M.; Meadowcroft, D.B.; Oakey, J.; and Simms, N.J., *Proc. 4th Berkeley Conf. Corrosion-Erosion-Wear of Materials at Elevated Temperatures* (Houston), A.V. Levy, Ed., NACE, 1991, p 23-1 to 23-12

36. Wang, B.Q.; Geng, G.Q.; and Levy, A.V., *Proc. 8th Int. Conf. Wear of Materials* (Orlando), 7-11 April, 1991, K.C. Ludema and R.G. Bayer, Ed., ASME, 1991, p 703-707

37. Bellman, R., and Levy, A.V., *Proc. Int. Conf. on Wear of Materials*, S.K. Rhee, A.W. Ruff, and K.C. Ludema, Ed., ASME, 1981, p 564-576

38. Hutchings, I.M., and Levy, A.V., *Wear*, Vol 131 (No. 1), 1989, p 105-121

39. Foley, T., and Levy, A., *Wear*, Vol 91 (No. 1), 1983, p 45-64

40. Robert, E., Reed-Hill, *Physical Metallurgy Principles*", 2nd ed., Van Nostrand Company, 1973, p 796-797

41. *Metals Handbook*, 9th ed., Vol 1, American Society for Metals, 1978, p 683

# Chapter 7

# Erosion of Protective Coatings

## Erosion-Corrosion of Thermal Spray Coatings[1]

Thermal spray coatings are being used on structural steels in energy conversion and utilization systems to prevent surface degradation by corrosion, erosion, or combinations of these mechanisms. The ability of the coatings to protect base materials against erosion-corrosion is related to their composition and processing, and especially to the form and structure of the coating. Generally, the hardness of the coatings does not directly relate to their erosion resistance (Ref 2-4). However, some data in the literature (Ref 5, 6) suggests that the relative hardness values of the erodent particles and the target material might play a role in the erosion behavior of brittle materials. It was reported that decreasing the ratio of the target material hardness to the hardness of the impacting particles resulted in an increase in the erosion rates, mainly when the ratio was in the range of unity.

In the work presented in this chapter, a number of metallic and hard material coatings designed to protect metal surfaces against erosion-corrosion were applied to 1018 steel by several spray processes. The erosion behavior of the coatings was determined at the standard test conditions listed in Chapter 6 that are used to determine the erosion-corrosion behavior of heat exchanger tubing steels. The test conditions approximately simulated the erosion-corrosion environment in the convection pass, superheater region of a CFBC. Two different erodents were used. All the

**Table 7.1  Coating parameters**

| Coating designation | Coating process | Coating material | Eroded by bed material No. 2 | | Eroded by $SiO_2$ | |
| --- | --- | --- | --- | --- | --- | --- |
| | | | Coating thickness(a), mm | Surface hardness, HV 100 | Coating thickness(a), mm | Surface hardness, HV 100 |
| 405 + 420 | Wire bonding and wire thermal spray | 420 stainless steel | 0.57 | 442 | 0.44 | 425 |
| JK420 | Hypersonic spray | 420 stainless steel | 0.21 | 455 | 0.07 | 439 |
| JK112 | Hypersonic spray | WC + 12% Co | 0.34 | 948 | 0.18 | 944 |
| Metco 74-SF | Plasma spray | WC + 12% Co | 0.10 | 780 | 0.06 | 739 |
| Metco 76F-NS (No. 1) | Plasma spray | WC + 18% Co | 0.15 | 997 | 0.10 | 992 |
| Metco 76F-NS (No. 2) | Plasma spray | WC + 18% Co | ... | ... | 0.17 | 1335 |
| SMI-28 | Thermal spray | WC + CrNiCo | 1.0 | 1011 | 0.50 | 936 |
| Metco 465-NS | Plasma spray | Fe-Cr-Al-Mo(b) | ... | ... | 0.37 | 408 |
| Metco 468-NS | Plasma spray | Ni-Cr-Al-Co(c) | 0.82 | 360 | 0.44 | 354 |

(a) The coating thickness were measured after polishing. (b) 27.5% Cr, 6% Al, 2% Mo, balance Fe. (c) 26.5% Cr, 7% Al, 3.5% Co, 1.0% $Y_2O_3$, balance Ni coatings were commercially available coatings applied by industrial thermal spray companies. The relationship between coating composition, morphology, and the material wastage was determined. Also the role of the ratio of the target hardness to the hardness of the erodent particles in the erosion-corrosion behavior of the coatings was studied.

### Test Conditions

The target materials for the erosion-corrosion tests were nine thermal spray coatings on mild steel. The characteristics of the coatings are listed in Table 7.1. The test surfaces of the coatings were polished to 600 grit prior to testing. The thicknesses of the coatings listed in Table 7.1 were measured after polishing.

The tests were carried out in the room-temperature and elevated-temperature, nozzle testers described earlier (Ref 7, 8). Air was utilized as the carrier fluid for the particles, creating a generally oxidizing atmosphere at elevated temperature. Two different erodents were used; 250 μm

CFBC bed material number 2 and 250 μm $SiO_2$-fused quartz. The CFBC bed material erodent had an angular shape and consisted primarily of $SiO_2$ and CaO, as determined by x-ray diffraction analysis. The detailed test conditions are listed at the bottom of Tables 7.2 and 7.3. The material wastage of the specimens was determined by microscopic dimensional measurements to determine the thickness changes of the specimens in the central part of the erosion zone, as described in Chapter 6.

### Erosion-Corrosion Rates

The material wastage of specimens eroded-corroded by the different erodent particles are listed in Table 7.2. For comparison, the wastage of the 1018 mild steel substrate (Ref 9) is also listed in Table 7.2. From Table 7.2 it can be seen that all the coatings tested, using both erodents, had lower material wastage than bare 1018 steel. Other, diffusion type coatings on mild steel, including chromizing and "Extendalloy" proprietary coating also significantly decreased metal degradation on heat exchanger tubes in atmos-

**Table 7.2   Material wastage of coatings**

| Coating | Eroded by bed material No. 2 | | Eroded by $SiO_2$ | |
| | Coating hardness/ erodent hardness | Thickness loss, μm | Coating hardness/ erodent hardness | Thickness loss, μm |
| --- | --- | --- | --- | --- |
| Metco 76F-NS, No. 2 | ... | ... | 1.00 | 2.5 |
| Metco 76F-NS, No. 1 | 1.43 | (a) | 0.75 | 2.7 |
| JK112 | 1.36 | 3.7 | 0.71 | 3.3 |
| SMI-28 | 1.45 | 1.8 | 0.70 | 3.4 |
| Metco 74 SF | 1.12 | 3.5 | 0.56 | 3.9 |
| JK420 | 0.65 | 4.0 | 0.33 | 5.3 |
| 405 + 420 | 0.63 | 3.2 | 0.32 | 5.7 |
| Metco 465-NS | ... | ... | 0.31 | 6.8 |
| Metco 468-NS | 0.51 | 4.5 | 0.27 | 7.8 |
| 1018 steel | ... | 6.6 | ... | 12.3 |

Test conditions: $T$ = 450 °C, $V$ = 20 m/s, $\alpha$ = 30°, $t$ = 5 h, loading 375 g. Hardnesses: $SiO_2$, 1329 HV; CFBC bed material No. 2, sand (76%) 881 HV, CaO (24%) 117 HV. (a) After test, coating broke away

**Table 7.3   Material wastage of three coatings at different test conditions**

| Designation | Thickness loss, μm | | | | | | | |
| | Eroded at room temperature | | | | Eroded at 450 °C | | | |
| | $V$ = 10 m/s | | $V$ = 20 m/s | | $V$ = 10 m/s | | $V$ = 20 m/s | |
| | $\alpha$ = 30° | $\alpha$ = 90° | $\alpha$ = 30° | $\alpha$ = 90° | $\alpha$ = 30° | $\alpha$ = 90° | $\alpha$ = 30° | $\alpha$ = 90° |
| --- | --- | --- | --- | --- | --- | --- | --- | --- |
| Metco 76F-NS (No. 1) | 1.4 | 7.2 | 2.6 | 9.4 | | 7.5 | 2.7 | 9.8 |
| SMI-28 | 2.4 | 2.8 | 3.3 | 6.3 | ... | | 3.4 | 10.1 |
| Metco 468 | 1.8 | 1.3 | 1.8 | 1.7 | 8.1 | | 7.8 | 11.6 |

Erodent particle, 250 μm $SiO_2$; $t$ = 5 h ($V$ = 20 m/s) or 8 h ($V$ = 10 m/s); loading, 375 g ($V$ = 20 m/s) or 600 g ($V$ = 10 m/s)

phere fluidized bed combustor service (Ref 10). Therefore, all the coatings tested were protective and can improve the erosion-corrosion behavior of 1018 mild steel tubing. Generally the $SiO_2$-fused quartz was more erosive than the CFBC bed material.

When the CFBC bed material was the erodent, the hardness of the coatings had no direct relationship to material wastage. The ductile, softer coatings had the same or somewhat lower material wastage than the harder tungsten carbide-cobalt coatings. When the $SiO_2$-fused quartz was the erodent, the hardness of the coatings had an inverse relationship to material wastage. The ratio of the coating hardness to erodent hardness versus the material wastage is plotted in Fig. 7.1. It can be seen that there are two ranges in the curve. Above a ratio of 0.56, the curve is smooth and shows a slight reduction in material wastage with an increase in the coating-erodent hardness ratio. Below a ratio of 0.33, the curve is a steep straight line showing a dramatic reduction in material wastage with an increase in the ratio. The former range corresponds to the hard material coatings, and the latter range covers the softer, ductile metal coatings; see Table 7.2.

Reference 6 reported that when the ratio of the target hardness to the hardness of the erodent particle approached unity, there was a sharp transition in erosion rates. In that investigation, target samples were brittle materials and the relationship between the hardness ratio and the erosion rate resulted from tests in which the same target material was eroded using erodents with different hardnesses. In this investigation, different coatings with different hardnesses were the target materials that were eroded using the same erodent. When the $SiO_2$-fused quartz was the only erodent used, there was a different correlation between the hardness ratio and the material wastage. A sharp transition occurred, but at a ratio of 0.4, not unity.

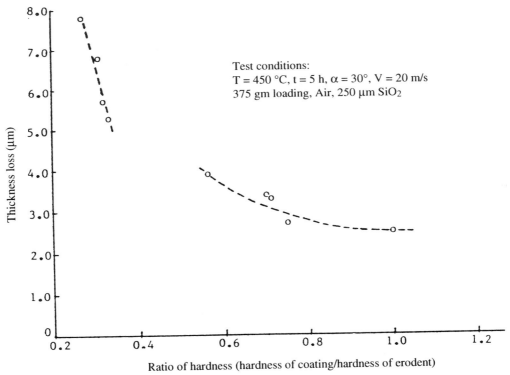

Test conditions:
T = 450 °C, t = 5 h, α = 30°, V = 20 m/s
375 gm loading, Air, 250 μm $SiO_2$

**Fig. 7.1** Relationship between the ratio of the coating hardness to the erodent hardness and the material wastage

As mentioned above, when CFBC bed material was the erodent, there was no correlation at all. The CFBC bed material contained CaO, a soft, weak constituent that is prone to shattering and smearing upon impact, which results in covering the surface of the target with a layer of material. Therefore, when this kind of erodent particle impacted the surface of the target, the damage pattern was no longer based on the indentation fracture behavior of the target material (Ref 11, 12).

From Table 7.2 it can also be seen that the coatings had different relative material thickness losses with each of the two erodents used. The WC-(Cr, Ni, Co) coating, SMI-28, had the lowest metal wastage and the WC-Co, Metco 76F NS coating, had the highest material wastage when CFBC-bed material was the erodent. When $SiO_2$-fused quartz was the erodent, the Metco 76F NS coating had the lowest material wastage.

In other investigations (Ref 2), it has been observed that the straight erosion behavior of coatings tested at room and elevated temperature was very dependent on the shapes and properties of the erodent particles. Different erosion behavior occurred for the same material or coating when it was impacted by different particles and the ranking of materials tested varied when different erodent particles were used. The data in Table 7.2 are from specimens tested at 450 °C. X-ray diffraction analysis determined that essentially no oxidation of the coatings tested occurred, except for the 420 stainless steel coatings. Therefore, the surface was only subjected to elevated-temperature erosion without oxidation (see Chapter 5).

## Effect of Test Variables on Material Wastage

Three thermal spray coatings, the two best-performing coatings and the worst-performing coating listed in Table 7.2, were eroded with 250 μm $SiO_2$-fused quartz at room temperature and 450 °C at impact angles of 30 and 90° and particle velocities of 10 and 20 m/s. The material wastages of the three coatings at different test conditions are listed in Table 7.3.

**Temperature.** When the test temperature was increased from room temperature to 450 °C, the material wastage of the two best-performing coatings, Metco 76F NS (No. 1) and SMI-28, increased only a small amount. However, the worst-performing coating at 450 °C, Metco 468, had the lowest material wastage of the three at room temperature. Increasing the test temperature from 25 to 450 °C greatly increased the material wastage of the Metco 468 coating at impact angles of both 30 and 90°, and at particle velocities of both 10 and 20 m/s.

**Impact Angle.** From Table 7.3, it can be seen that the material wastages of the Metco 468 coating at room temperature for all of the test conditions were nearly the same. However, the other two coatings tested had erosion behavior that was typical of brittle materials at both test temperatures and both particle velocities with greater material loss at an impact angle of 90° than at 30°. Even the Metco 468 metallic coating at 450 °C had the erosion behavior of a brittle material. The data listed in Ref 13 indicated that both the elongation and the strength of some Ni-Cr-Al alloys dramatically decreased at certain elevated temperatures. The Metco 468 coating may be one of these alloys. It is postulated that the Metco 468 coating, which is a Ni-Cr-Al alloy, had the best ductility at room temperature, which accounted for its low material wastage. Upon increasing the test temperature to 450 °C, the ductility of this coating was dramatically reduced, resulting in the higher material wastage and the erosion behavior of a brittle material.

**Particle Velocity.** In Table 7.3 it can be seen that when the particle velocity was increased from 10 to 20 m/s at both test temperatures the material losses increased for the three coatings even though the erodent loading was reduced from 600 to 375 g and the test time was reduced from 8 to 5 h. The effect of particle velocity on changing the thickness loss was less at room temperature for the Metco 468 coating than for the others.

## Metallographic Analysis

The erosion material wastage rates correlated well with the composition and morphology of the coatings. Figure 7.2 shows that the flame spray 420 stainless steel eroded by the cracking and chipping of a thin, intermittent Fe + Cr oxide scale layer, which was corroborated by x-ray diffraction analysis. The scale layer had bed material, calcium and silicon, embedded in it. The coating was quite dense with only a few small pores present in the area shown. Contrast this type of porosity with that shown in Fig. 7.3 for the hypersonically sprayed 420 stainless steel. The presence of more and larger pores resulted in a 25% higher material wastage when CFBC bed material was the erodent, 4.0 μm, compared to 3.2 μm as seen in Table 7.2.

When $SiO_2$ was the erodent, material loss of the metal coatings was by direct erosion of the metal by the platelet mechanism of erosion, as

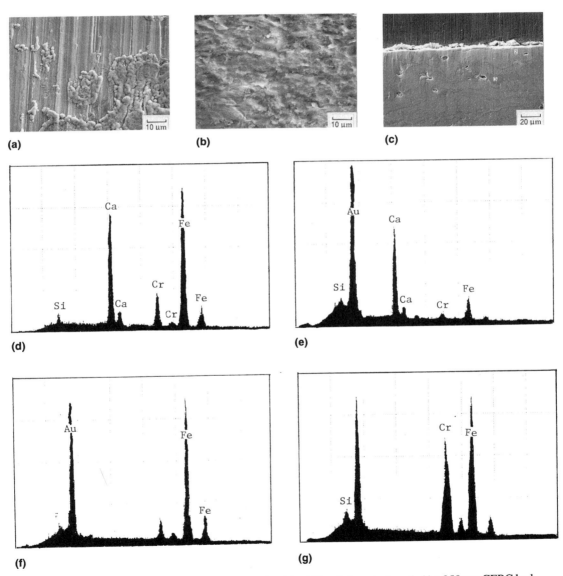

**Fig. 7.2** Surface and cross section of flame sprayed 405 + 420 stainless steel eroded by 250 μm CFBC bed material No. 2. $T = 450$ °C, $V = 20$ m/s, $\alpha = 30°$, $t = 5$ h (375 g). (a) Uneroded surface. (b) Eroded surface. (c) Cross section. (d) Composition of eroded surface. (e) Composition of cross section at 1. (f) Composition of cross section at 2. (g) Composition of cross section at 3

can be seen in Fig. 7.4. The erodent particles did not embed to form a layer as occurred when CFBC bed material was the erodent. The material wastage was greater when $SiO_2$ was the erodent for all the coatings (see Table 7.2) indicating that the scale layer of bed material and iron oxide that formed when the erodent was CFBC bed material provided some degree of protection to the coating. The bed material erodent caused the forma-

tion of the protective layer, insulating the substrate from the direct erosion process, as was discussed in detail in Chapter 6. Hence, the carbides and metal alloys, both being protected, had similar material wastage except for the SMI-28 coating. When $SiO_2$ was used as the erodent and the coating was being directly eroded, the high WC content coatings had lower thickness losses than the softer metals.

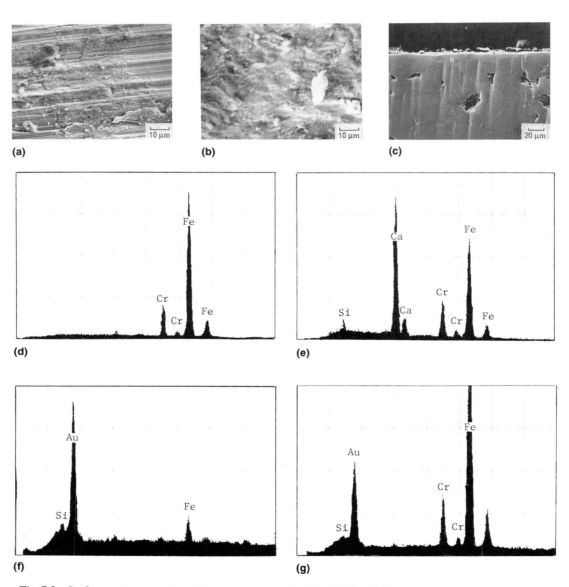

**Fig. 7.3** Surface and cross section of hypersonic sprayed (JK420) 420 stainless steel eroded by CFBC bed material No. 2. $T = 450\ °C$, $V = 20$ m/s, $\alpha = 30°$, $t = 5$ h (375 g). (a) Uneroded surface. (b) Eroded surface. (c) Cross section. (d) Composition of uneroded surface. (e) Composition of eroded surface. (f) Composition of cross section at 1. (g) Composition of cross section at 2

The WC + Co coatings eroded by a cracking and chipping mechanism, as can be seen in Fig. 7.5. The excess loss of the Metco 76F-NS (No. 1) coating eroded by CFBC bed material was due to a defective coating-base metal bond. Some areas of this coating broke away after the test. The reason for the low metal wastage of the SMI-28 coating eroded by CFBC bed material can be seen in Fig. 7.6. The coating was dense, with relatively few, small pores. Compare this coating in Table 7.2 with the more porous, hypersonic spray JK112 coating whose material loss was twice as

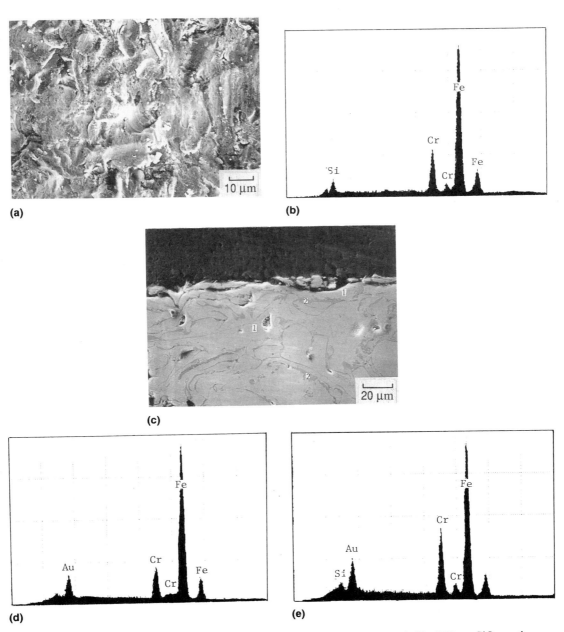

**(a)**

**(b)**

**(c)**

**(d)**

**(e)**

**Fig. 7.4**   Surface and cross section of flame sprayed 405 + 420 stainless steel eroded by 250 μm $SiO_2$ particles. $T = 450$ °C, $V = 20$ m/s, $\alpha = 30°$, $t = 5$ h (375 g). (a) Surface. (b) Composition of surface. (c) Cross section. (d) Composition of cross section at 1. (e) Composition of cross section at 2

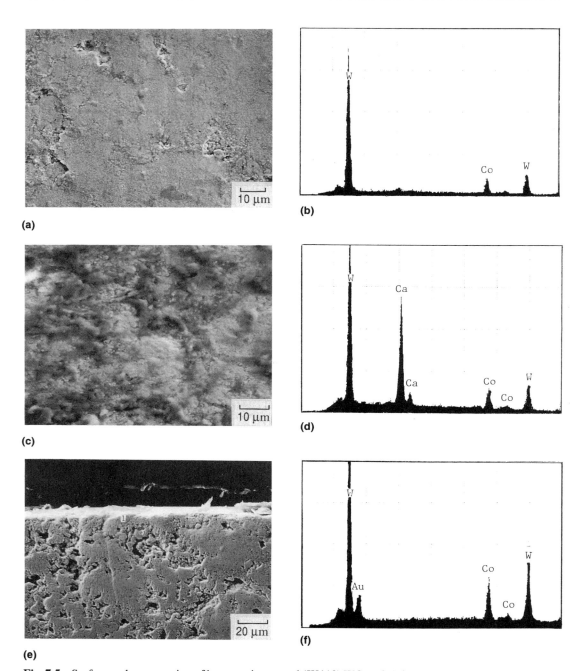

**Fig. 7.5** Surface and cross section of hypersonic sprayed (JK112) WC + 12% Co eroded by 250 μm CFBC bed material No. 2. $T = 450$ °C, $V = 20$ m/s, $\alpha = 30°$, $t = 5$ h (375 g). (a) Uneroded surface. (b) Composition of uneroded surface. (c) Eroded surface. (d) Composition of eroded surface. (e) Cross section. (f) Composition of cross section at 1

great. In this case, the significant difference in the nature of the porosity of the coating overcame the protective effect of the erodent-containing layer on the coating surface.

Figure 7.7 shows the cracked and chipped surface of the Metco 76F-NS (No. 2) coating eroded by $SiO_2$. The smooth character of the eroded surface indicated that only very small

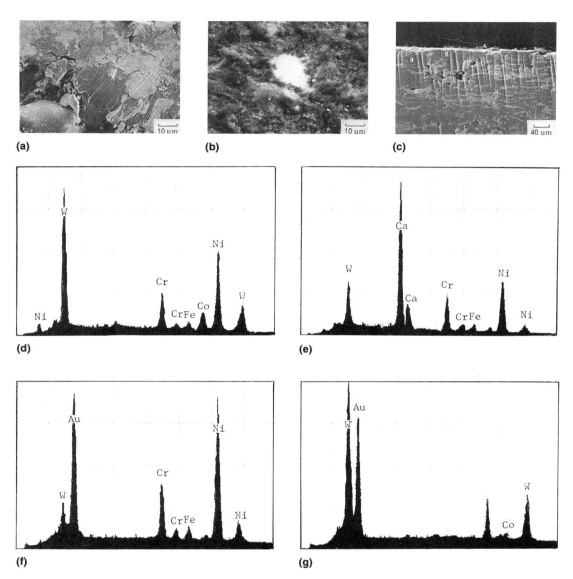

**Fig. 7.6** Surface and cross section of SMI-28 coating eroded by 250 μm CFBC bed material No. 2. $T = 450$ °C, $V = 20$ m/s, $\alpha = 30°$, $t = 5$ h (375 g). (a) Uneroded surface. (b) Eroded surface. (c) Cross section. (d) Composition of uneroded surface. (e) Composition of eroded surface. (f) Composition of cross section at 1. (g) Composition of cross section at 2

chips could be eroded, accounting for the low 2.5 μm, material thickness loss. No $SiO_2$ was embedded in the surface of the coating.

The Fe-Cr-Al-Mo coating (Metco 465) that had high metal wastage in the 5 h test using $SiO_2$ erodent is shown in Fig. 7.8. The Ni-Cr-Al-Co Metco 468 coating had the same morphology as the coating shown in this figure. It lost material

by the direct erosion of the metal with no protection from any kind of layer on its surface.

## Erosion of Protective Coatings for Steam Turbine Components

The erosion of ferritic stainless steels in elevated-temperature service in the nozzle and

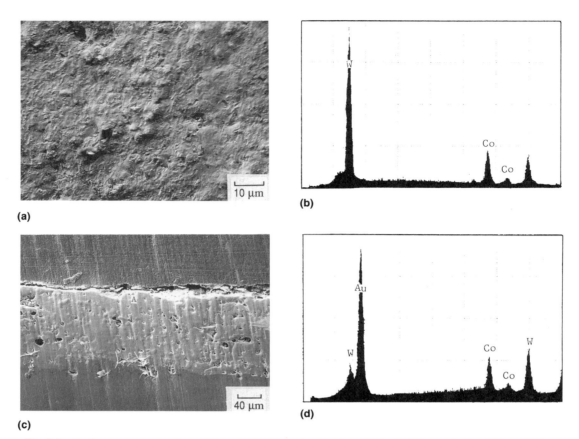

**Fig. 7.7**  Surface and cross section of Metco 76F-NS coating No. 2 on WC + 18% Co eroded by 250 μm SiO₂. $T = 450$ °C, $V = 20$ m/s, $\alpha = 30°$, $t = 5$ h (375 g). (a) Surface. (b) Composition of surface. (c) Cross section. (d) Composition of cross section at A

blades of steam turbines prompted a research and development program to protect the surfaces of the components by the application of thin protective coatings. A group of twelve coatings of various compositions and application methods were applied to types 403 and 422 stainless steel and, along with uncoated 403 and 422 were laboratory erosion tested (Ref 2). The test results of the coatings on 422 are listed in Table 7.4. The elevated-temperature test conditions were similar to service conditions except for the particle velocity, which was one-half the 300 m/s that occurs in service and the use of air as the carrier gas instead of dry steam. The tests were all carried out for 26 min to essentially steady-state erosion conditions. A distinct order of performance occurred for the coating materials. The order was related to the mechanism of erosion of each coating as it was affected by its grain size, porosity,

and the presence of cracks formed during processing.

### Test Conditions

All of the coatings tested were applied to 403 and 422 ferritic stainless steels by coating suppliers. Table 7.4 lists all of the coatings tested. Because the coatings were all proprietary, details regarding their precise composition and application process variables were not available. The coating suppliers were purposely not identified in the table.

The specimens tested varied somewhat in their base metal dimensions but were generally 5 cm long by 1.9 cm wide by 0.3 cm thick plus the thickness of the coating as indicated in Table 7.4. The coating surfaces were in their as-applied condition.

**Table 7.4  Erosion of coatings on 422 stainless steel**

| Ranking α=30° | Ranking α=90° | Material No. | Name | Application technique | Surface hardness, HV 500 | Density, g/cm³ | Thickness as-applied, μm | Thickness loss at α=30°, μm | Erosion at α=30° cm³/g ×10⁻⁴ No. 1 | No. 2 | Avg | Erosion at α=90° cm³/g ×10⁻⁴ No. 1 | No. 2 | Avg |
|---|---|---|---|---|---|---|---|---|---|---|---|---|---|---|
| 14 | 11 | 0 | 422SS | Bare | 275 | 7.8 | ... | 126 | 1.73 | 1.52 | 1.63 | ... | 0.65 | ... |
| 1 | 7 | 3 | Cr₃C₂ | Deton. gun | 1009 | 6.4 | 120 | 4.5 | 0.04 | 0.05 | 0.05 | ... | 0.29 | ... |
| 2 | 8 | 8 | WC-NiCrB | Clad | 982 | 11.5 | 130 | 5 | 0.05 | 0.06 | 0.06 | ... | 0.3 | ... |
| 3 | 3 | 12-A | FeB, Fe₂B | Diffused(b) | ... | 7.1 | 70 | ... | 0.073 | ... | ... | ... | 0.185 | 0.185 |
| 4 | 9 | 91A-2 | WC | CVD | 713 | ... | 60 | 7.5 | 0.09 | 0.09 | 0.09 | 0.34 | 0.31 | 0.33 |
| 5 | 10 | 4 | TiN | PVD | 516 | 5.2 | 15 | 8 | 0.08 | 0.10 | 0.09 | ... | 0.57 | ... |
| 6 | 2 | 91A-9 | WC | CVD | 795 | ... | 40 | 10 | 0.12 | ... | ... | 0.10 | ... | ... |
| 7 | 1 | 91A-10 | WC | CVD | 549 | ... | 43 | 12 | 0.14 | ... | ... | 0.06 | ... | ... |
| 8 | 4 | 2 | CrB | Diffused | 1283 | 7.1 | 90 | 16 | 0.18 | 0.19 | 0.18 | 0.19 | ... | ... |
| 9 | 5 | 12-B | FeB, Fe₂B | Diffused(c) | ... | 7.1 | 70 | ... | 0.19 | ... | ... | 0.196 | ... | ... |
| 10 | 6 | 91A-3 | WC | CVD | 644 | ... | 110 | 16 | 0.20 | 0.20 | 0.20 | 0.29 | 0.27 | 0.28 |
| 11 | 13 | 6 | WC-Co | Plasma spray | 927 | 13.8 | 120 | 19 | 0.21 | 0.23 | 0.22 | ... | 0.83 | ... |
| 12 | 12 | 5 | NiCrBC | Plasma spray | 1006 | 12.5 | 110 | 54 | 0.65 | 0.64 | 0.65 | ... | 0.80 | ... |
| 13 | 14 | 1 | 422SS | Nitrided | 1027 | 7.1 | 250 | 62 | 0.72 | 0.75 | 0.73 | ... | 1.18 | ... |
| Wore through | ... | 10 | Cr₃C₂ | Sputtered | 613 | 7.1 | 25 | 25 | ... | ... | ... | ... | ... | ... |
| Wore through | ... | 11 | Cr₃C₂ Trib 800 | Sputtered | 615 | 8.6 | 20 | 20 | ... | ... | ... | ... | ... | ... |
| Wore through | ... | 7 | Nickel + P | Electroless nickel | 617 | 7.9 | 80 | 80 | ... | ... | ... | ... | ... | ... |

Test conditions: $V = 150$ m/s (500 ft/s), $T = 538$ °C (1000 °F), erodent: 25 g of 74 μm diam chromite in air carrier gas. (a) Corrected for coating density. (b) Faster cooling rate. (c) Standard cooling rate

**Table 7.5  Test conditions for hard material coating systems**

| | |
|---|---|
| Temperature, °C (°F) | 538 (1000) |
| Particle velocity, m/s (ft/s) | 150 (500) |
| Impingement angle | 30°, 90° |
| Carrier gas | Undried air |
| Particle composition, chromite wt% | 38% Cr₂O₃, 24% Fe₂O₃, 16% MgO, 15% Al₂O₃, 7% SiO₂ |
| Particle size, μm | 74 average |
| Particle load, g | 25 |
| Test time, min | 26.5 |

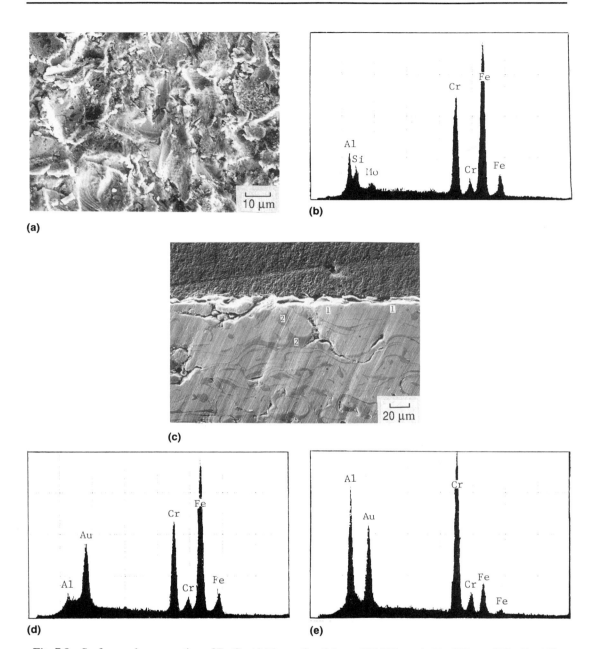

**Fig. 7.8** Surface and cross section of Fe-Cr-Al-Mo coating (Metco 465-NS) eroded by 250 μm SiO$_2$. $T$ = 450 °C, $V$ = 20 m/s, $\alpha$ = 30°, $t$ = 5 h (375 g). (a) Surface. (b) Composition of surface. (c) Cross section. (d) Composition of cross section at 1. (e) Composition of cross section at 2

The test conditions are listed in Table 7.5. The elevated temperature erosion nozzle tester was used that is described in Chapter 5. Erosion rates were determined by periodic weight loss measurements using an analytical balance that measured specimen weight to 0.1 mg. After testing, the specimen surfaces were observed in a scanning electron microscope with energy dispersive spectroscopy (EDS).

## Erosion Rates

Table 7.4 is a compilation of the erosion data for all of the materials tested on 422 stainless steel

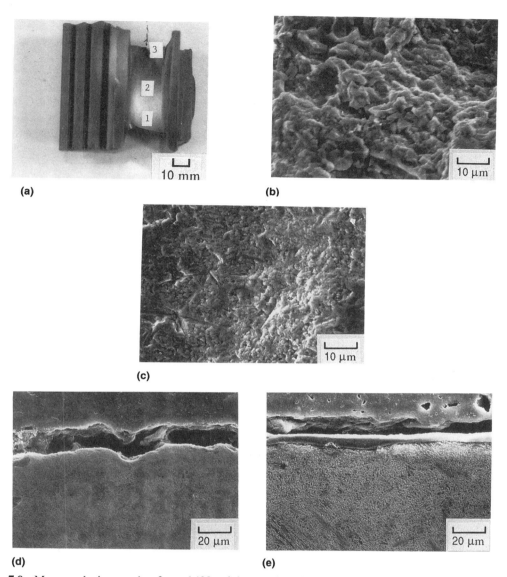

**Fig. 7.9**  Macro and micrographs of tested 422 stainless steel turbine blade. (a) Concave side. (b) Surface at position 1. (c) Surface at position 2. (d) Cross section at position 1. (e) Cross section at position 2

along with some characterization information. The results of the coatings tested on stainless steel 403 are contained in Ref 2. The materials are listed in the order of their performance in the 30° impact test except that bare 422 stainless steel is listed first so the improvements in erosion behavior due to the various coatings can be easily compared. The consistency of the test machine can be seen in the 30° impact data. Some of the materials were only tested once at a 90° impact. The rankings differed between the 30 and 90°

impact angle tests. Three materials had the coatings completely removed in the 30° impact tests. Hardness of the coating does not have a direct effect on the erosion of the coatings. In general, the rankings of the coatings on each of the two steel substrates were comparable. The coatings are discussed in the order that they appear in Table 7.4.

**Uncoated 422 Stainless Steel.** Figure 7.9 shows a macrograph of an in-service tested, uncoated steam turbine blade and micrographs of the surface and cross section at positions 1 and 2. There was iron oxide scale with a small amount of chromium dissolved in it on all three surface areas examined, as determined by EDS peak analysis. The surface at position 1 shows the typical cracked and chipped morphology of a brittle material being eroded at a steep impact angle. The cross section shows that the oxide broke away from the metal surface when it was being cut for mounting, which occurs when the scale is thick and a fine diamond cutting wheel is not used, which was the case here. Based on the work discussed in Chapter 5, the thickness of the scale could have been enhanced by the impacting particles (Ref 14).

The surface at position 2 is typical of a thin oxide being eroded at a more shallow angle. The narrow, long gouges that can be seen on the surface indicate that the thin oxide was locally plastically deformed into the base metal, which was also being deformed. The scale was too thin to isolate the base metal from indentation deformation by sharp protrusions on the erodent particles. The thin cross section of oxide on the 422 stainless steel can be seen in Fig. 7.9(e). Position 3 had some areas where no scale formed and erosion occurred by the platelet mechanism on bare, ductile metal.

The surface of a 422 stainless steel specimen eroded in the laboratory nozzle tester at a 30° impact angle is shown in Fig. 7.10. In the low magnification micrograph (Fig. 10a), it can be seen that rippling of the surface had occurred. Rippling occurs on some ductile metals eroded at shallow impact angles. The higher magnification photo (Fig. 7.10b) of the surface shows a fine distribution of platelets, shallow craters, and nar-

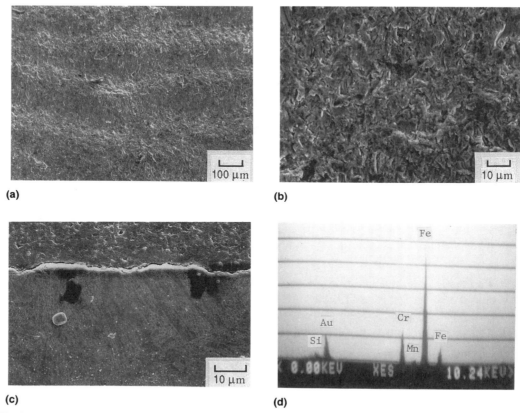

(a)  (b)  (c)  (d)

**Fig. 7.10**  Surface and cross section of uncoated 422 stainless steel eroded at $\alpha = 30°$. 74 μm mixed oxide erodent. $T = 538$ °C, $V = 150$ m/s, $t = 26.5$ min (75 g). (a) Eroded center. (b) Eroded center (higher magnification). (c) Center cross section. (d) Composition of cross section

row gouges. The size of the structural elements can be related to the fine microstructure of the base metal. There is no evidence of oxide scale on the surface. The EDS peak analysis is typical of a 400 series stainless steel. The cross section shows some fine platelets that have been formed by the erosion process, which were also seen on the tested blade at position 3, Fig. 7.9(a).

The surface of the 422 stainless steel specimen tested at a 90° impact angle indicates that considerable embedding of the erodent particles had occurred at the elevated test temperature. EDS peak analysis indicates that the constituents that were in the erodent were contained in the target metal surface. The lower erosion rate at a 90° impact listed in Table 7.4 could, in part, be due

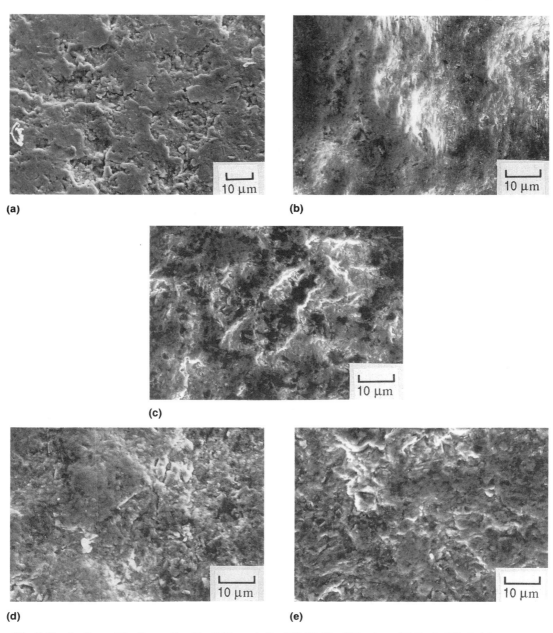

**Fig. 7.11** Surface of $Cr_3C_2$, coating No. 3 (D-gun). $T = 538$ °C, $V = 150$ m/s, $t = 26.5$ min (75 g). (a) Uneroded surface. (b) Center ($\alpha = 30°$). (c) Outer zone ($\alpha = 30°$). (d) Center ($\alpha = 90°$). (e) Outer zone ($\alpha = 90°$)

to some level of protection that was provided by the embedded particles.

**Cr₃C₂ Coating No. 3.** The surfaces of the best-performing coating from the 30° and 90° impact tests are shown in Fig. 7.11. The uneroded D-gun, sprayed material consists of very fine-grain particles with very little porosity. The surface was ground after deposition. When the coating was eroded at a 30° impact angle, it was further smoothed and some of the voids were filled with eroding debris. This can be seen in the surface at the center of the eroded area (Fig. 7.11b). In the outer zone of the eroded area (Fig. 7.11c), fewer particle impacts occurred and the coating surface was much coarser. Some of the small voids in the as-sprayed coating have been opened up (black areas), and the smooth, ground surfaces have been roughened.

At 90° impact the eroded surface at the center is rougher than occurred in the 30° test. The fine-grain structure of the coating can be readily seen in both the center (Fig. 7.11d) and outer-zone (Fig. 7.11e) micrographs, which had basically the same morphology and were very similar to the surface of the uneroded coating.

The excellent performance of this coating is directly related to the fine-grain, low-porosity morphology of the coating. It was removed by cracking and chipping off of very small pieces whose size was determined by the basic grain size of the coating (see Chapter 2). In the erosion of brittle materials, the finer the grain size and the lower the porosity, the lower is the erosion rate because only small pieces can be chipped off by particle impacts (Ref 15). The EDS peak analysis of this coating shows that the surface consisted primarily of Cr₃C₂ with a small amount of erodent embedded in the pores (Cr, Mg, Al, Si, Fe).

**WC + NiCrB Coating No. 8.** The surface of the uneroded, clad coating in Fig. 7.12 shows a fine grain structure that was ground after cladding. The cross section (Fig. 12c) and the EDS peak analysis (Fig. 7.12e) indicates that the material consisted of grains of WC of various sizes in a matrix of Ni-Cr-B alloy. A few voids of reasonable size were present as were some cracks. The surface and cross section mi-

crographs of the center zone from the 90° impact test shows that the matrix was preferentially eroded, leaving higher WC particles. The outer zone appears to have the voids filled, probably with fine debris.

The shallow, 30° impact test specimen surface indicated that both the WC particles and the Ni-Cr-B matrix alloy eroded at nearly the same rate. The WC appeared to provide some protection for the matrix as the erosion rate at 30° impact was much lower than at 90° impact (see Table 7.4). As in the case of the sprayed Cr₃C₂, coating No. 3, the nearly 100% dense structure of the coating resulted in a low erosion rate. The widely dispersed voids and cracks did not result in a major reduction in the performance of the coating.

**WC-CM500L Coating No. 9, 1A-2.** The CM500L, chemical vapor deposited (CVD) coating was tested in the heat-treated condition. The uneroded material was dense and had a very small grain size that is known from previous work to be about 100 Å. There was some evidence that a second deposition to fill cracks left in the first deposited layer had occurred. This type of dense, very small grain size microstructure erodes by cracking and chipping off of very small particles; hence the low material loss rates listed in Table 7.4.

**TiN Coating No. 4.** This physical vapor deposited (PVD) coating was only 15 μm thick. At 30° impact, its fine grain, dense morphology resisted the impacting particles well enough to leave a thin coating layer at the end of the short time test. The eroded surface had the fine-grain, smooth surface that occurred when material loss was caused by cracking and chipping of very fine pieces. The EDS peak analysis indicated that the TiN completely covered the surface. The relatively low hardness of this coating indicates that it is the fine grain structure and high density that determined the good erosion behavior and not the hardness. The thin coating on the surface of the 90° impact specimen completely eroded off, and the final eroding surface was the base metal. The complete loss of the coating may be indicative of the brittleness of the TiN. If this is the case, a thicker coating may or may not have been successful.

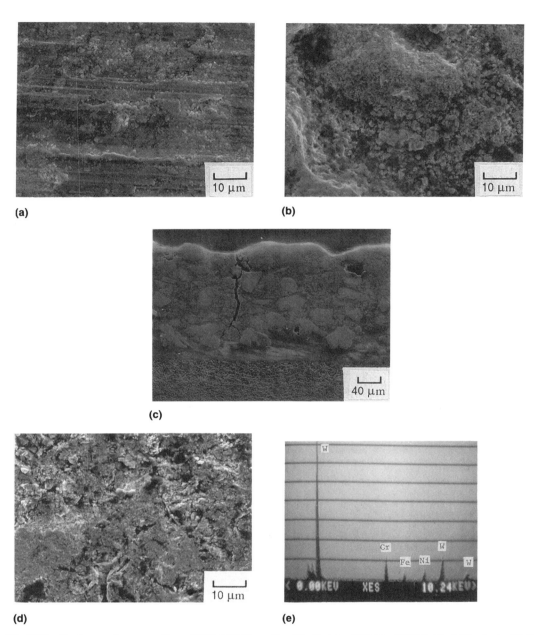

**Fig. 7.12** Surface and cross section of WC-NiCrB, coating No. 8 at $\alpha = 90°$. $T = 538\ °C$, $V = 150\ m/s$, $t = 26.5$ min (75 g). (a) Uneroded edge. (b) Center. (c) Center cross section. (d) Outer zone. (e) Center composition

**WC-CM500L Coatings, No. 9: 1A-9 and 1A-10.** The 1A-9 and 1A-10 versions of the CVD WC coating were essentially the same in their morphologies and in their erosion rates. Figure 7.13 shows the representative 1A-10 specimen surface and cross section. It had a smooth surface texture and very fine grain size, from which material had been generally removed by cracking and chipping of very small pieces. This version of the CM500L coating was deposited with a much finer surface texture than the CM500L coating No. 9, 1A-2 by modifying the deposition conditions. However, there is evidence that the significant cracking of the material, which can be seen in both the surface and cross section micrographs, caused periodic erosion loss of large

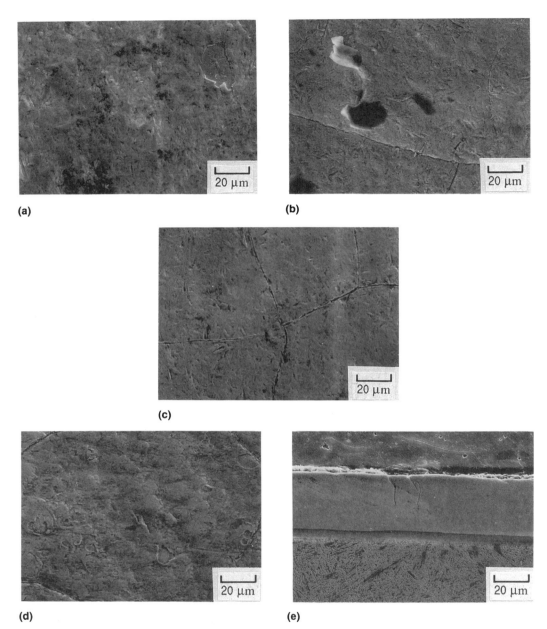

**Fig. 7.13** Surface and cross section of CM500L, coating No. 9 (chemically vapor deposited WC), 1A-10. $T =$ 538 °C, $V = 150$ m/s, $t = 26.5$ min (75 g). (a) Center ($\alpha = 30°$). (b) Halo zone ($\alpha = 30°$). (c) Center ($\alpha = 90°$). (d) Halo zone ($\alpha = 90°$). (e) Center cross section ($\alpha = 90°$)

pieces. Such a loss was documented in the cross section of coating CM500L-1A-3, which is discussed later in this chapter.

Thus, the very fine as-deposited grain size was compromised by the cracks and the resultant erosion rate was an intermediate one. Modification of the processing conditions had a major effect on the performance of the resulting coating. This indicates the potential of coating performance improvement on the one hand, and the need for tight process quality control on the other.

**CrB Coating No. 2.** Figure 7.14 shows the 30° impact angle specimen microstructure of the boron diffused coating that appeared to have CrB

more severely cracked network. The coating was quite thin, as deposited. No coating remained at the end of the test.

**Ni + P Coating No. 7.** This electroless nickel coating had large pores and a layered microstructure. Both conditions made it highly susceptible to erosion. The eroded surface had no coating remaining on it at the end of the test.

## Discussion

This series of coatings had a spectrum of compositions and microstructures from almost every method of applying a coating to a metal substrate that makes it an ideal vehicle to gain a basic understanding of how and why coatings behave over a wide range of conditions. There is a consistent pattern in their behavior. This pattern of behavior of the coatings related more to their grain size and defect structure than to their composition or hardness. The finer the grain size and the more dense the coating, the lower was the erosion rate. When the deposition process resulted in cracks in the coating, the larger and more profuse the cracks were, the higher was the erosion rate.

The reason for this correlation is the basic mechanism of particle removal. The size of the chips that formed when the impacting particles struck the generally brittle surface determined how much material could be removed by a given number of particle impacts in an area of the surface. Very fine grains resulted in small chips in the coatings and a resultant low loss rate. Larger grains and segments of coating between the pores (Ref 6) formed larger chips when impacted, with resultant larger loss rates. This behavior conforms to that described in Chapters 2 and 5.

When pre-existing cracks were present, such as the craze cracks that formed in several of the coatings tested, even larger pieces could be knocked off the surface by impacting particles with larger material loss rates. The three coatings which completely eroded off during the test all had extensive cracking in their as-deposited condition.

The overall hardness of the deposited or diffused coatings had relatively little effect on their ability to resist erosion. Hardness variations shown in Table 7.4 range from 516 to 1283 HV 500. Two of the better performing coatings (No. 4 and 9) had hardnesses at the lower end of the range while two of the poorer performing coatings had hardnesses greater than 1000 HV 500.

The erosion resistance of some of the coatings was markedly improved during the course of the test program by process modifications. The behavior of different versions of coatings 9 and 12, as listed in Table 7.4 indicated that considerable improvement may be possible by further modifications of the coatings. The process changes made primarily modified the crack structures that were present in the coatings. Changes in other aspects of the coatings such as composition and morphology could also result in performance improvements.

### References

1. Wang, B.Q.; Geng, G.Q.; Levy, A.V., Erosion-Corrosion of Thermal Spray Coatings, *Surf. Coat. Technol.*, Vol 43/44, 1990, p 859-874

2. Levy, A.V., and Wang, B.Q., Erosion of Hard Material Coating Systems, *Wear*, Vol 121 (No. 3), 1988, p 325-346

3. Levy, A.V., The Erosion-Corrosion Behavior of Protective Coatings, *Surf. Coat. Technol.*, Vol 36, 1988, p 387-406

4. Shui, Z.R.; Wang, B.Q.; and Levy, A.V., Erosion of Protective Coatings, *Surf. Coat. Technol.*, Vol 43/44, 1990, p 875-887

5. Scattergood, R.O., and Srinivasan, S., Erosion of Brittle Materials, *Proc. Conf. Tribological Mechanisms and Wear Problems in Materials*, Materials Week '87 (Cincinnati, OH), Oct 1987, ASM International, p 1-10

6. Srinivasan, S., and Scattergood, R.O., Effect of Erodent Hardness on Erosion of Brittle Materials, *Wear*, Vol 128, 1988, p 139-152

7. Levy, A.V., The Solid Particle Erosion Behavior of Steel as a Function of Microstructure, *Wear*, Vol 68 (No. 3), 1981, p 269-287

8. Levy, A.V.; Yan, J.; and Patterson, J., Elevated Temperature Erosion of Steels, *Wear*, Vol 108 (No. 1), 1986, p 43-60

9. Levy, A.V.; Wang, B.Q.; and Geng, G.Q., Characteristics of Erosion Particles in Fluidized Bed Combustors, *Proc. ASME 10th FBC Conf.* (San Francisco) May 1989, ASME, New York, 1989, p 945-951

10. Lewis, E.C., and Canonico, D.A., "Metal Wastage Experiences in BFBC Environments," Paper No. 20, Proc. NACE Conf. Corrosion Erosion-Wear of Materials at Elevated Temperatures, Berkeley, CA, 1990

11. Lawn, B.R., and Wilshaw, R., Indentation Fracture: Principles and Application, *J. Mater. Sci.*, Vol 10, 1975, p 1049

12. Evans, A.G., and Wilshaw, R., Quasi-Static Solid Particle Damage in Brittle Solids-I. Observations, Analysis and Implications, *Acta Metall.*, Vol 24, 1976, p 939

13. Brandes, E.A., Chapter 22, *Smithells Metals Reference Book*, 6th ed., Butterworths, London, 1983, p 71-72

14. Levy, A.V., and Man, Y.-F., Elevated Temperature Erosion Corrosion of 9Cr1Mo Steel, *Wear*, Vol 111 (No. 2), 1986, p 135-160

15. Levy, A.V.; Bakker, T.; Scholz, E.; and Aghazadeh, M., Erosion of Hard Metal Coatings, *Proc. High Temperature Protective Coatings Conf.*, The Metallurgical Society, AIME, 1983, p 339

16. Foley, T., and Levy, A.V., The Erosion of Heat Treated Steels, *Wear*, Vol 91 (No. 1), 1983, p 45

17. Davis, A.; Boone, D.; and Levy, A.V., Erosion of Ceramic Thermal Barrier Coatings, *Wear*, Vol 110 (No. 2), 1986, p 101-116

# Chapter 8
# Slurry Erosion

## Liquid-Solid Particle Slurry Erosion of Steels

The flow of liquids containing small solid particles through the passages of chemical process or transport systems causes erosion degradation of the containment metal surfaces. The degradation is similar in mechanism to that which occurs on ductile metals exposed to streams of solid particles in gases, see Chapter 2. In the relatively low velocity regimes in fossil-fuel combustion or conversion equipment, the knowledge of erosion mechanisms gained in either the gas or liquid fluid media can be used to understand the mechanism of metal removal in the other. Measuring the effects of several slurry test variables such as flow velocity, particle size, solids loading, and temperature on the erosion of steel alloys provides information to determine the behavior of steel containment materials in slurry flow service.

The particular type of service that was investigated and described in this chapter (Ref 1) related to the liquefaction of coal. The particles used in the tests were bituminous coal and the liquids were organic solvents of various compositions and viscosities. The resulting information is applicable to a wide range of slurries and their flow conditions.

### Experimental Conditions

One of the devices that was used to determine the effect of slurries of coal in an organic solvent on the surface material loss of a group of steel alloys that are commonly used in the flow passages of chemical process equipment was a slurry pot tester. The steels tested are listed in Table 8.1. The device simulates various aspects of the environment of the slurry flow in the region of a coal liquefaction system where the pulverized coal is mixed with the process solvent. The principal variables that affect the erosion process; velocity,

**Table 8.1   Steel compositions**

| Alloy designation | Typical composition, wt% | | | | | | |
| | Cr | Ni | Mo | Mn | Si | C | Fe |
|---|---|---|---|---|---|---|---|
| A53 (1018) | ... | ... | ... | 1.2 | ... | 0.20 | bal |
| A106 | ... | ... | ... | 1.0 | 0.1 | 0.20 | bal |
| 2.25Cr-1Mo | 2.25 | ... | 1 | 0.5 | 0.5 | 0.15 | bal |
| 5Cr-0.5Mo | 5 | ... | 0.5 | 0.5 | 0.5 | 0.15 | bal |
| 9Cr-1Mo | 9 | ... | 1 | 0.5 | 0.5 | 0.15 | bal |
| 410 | 12 | ... | ... | 1 | 1 | 0.15 | bal |
| 304 | 18 | 8 | ... | 2 | 1 | 0.08 | bal |
| 310 | 25 | 20 | ... | 2 | 1 | 0.25 | bal |
| 316 | 16 | 10 | 2.5 | 2 | 1 | 0.08 | bal |
| 321 | 17 | 9 | ... | 2 | 1 | 0.08 | bal |

impact angle, solids loading, particle size, and the combination of viscosity of the slurry and its temperature were simulated in the tests.

The slurry pot tester, in effect, directs a flow of coal particles in a liquid kerosene carrier at 3 mm diameter cylindrical specimens 5 cm long by rotating the vertically oriented specimens on the end of arms attached to a motor-driven rotating shaft in a controlled environment pot of slurry of 3 L capacity. Figures 8.1 and 8.2 show the exterior appearance of the slurry pot tester and its design cross section, respectively. A vortex is prevented by placing metal baffles around the inside periphery of the pot and settling of the coal

particles is prevented by a small propeller that is mounted near the bottom of the rotating shaft. The slurry pot operating temperature is controlled by cooling or heating from 25 to 250 °C.

Erosion behavior was determined by the weight loss of the cylindrical test specimens after rotating in slurry for incremental times up to 5 h. The angle of impact of the slurry particles on the test cylinder varies around the front half of the cylinder; so the weight loss is an integration of weight losses over the spectrum of impact angles from 0 to 90°. The reproducibility of the slurry pot tester is shown in Fig. 8.3 for two alloys, A53 mild steel and 304 stainless steel. Their overlap-

**Fig. 8.1**  Exterior view of slurry pot tester

ping erosion behavior is consistently observed over a large variation in test conditions.

### Test Results[Ref 2]

**Effect of Alloy Composition.** Figure 8.4 shows how a number of alloys eroded in a cumulative plot of erosion loss over a 2 h exposure period. Figure 8.5 shows the variation in erosion behavior of several austenitic stainless steels compared to A53 mild steel. It can be seen that the A53 and 304 stainless steel eroded similarly along with 321 stainless steel, which is a stabilized version of 304 stainless steel. A higher chromium content stainless steel, 310, eroded less, and a steel containing 2.5% Mo, 316 stainless steel, eroded the least. An attempt was made to correlate the erosion rates with some of the tensile strength and hardness properties and chromium content of the alloys tested. No correlation was found for the mechanical properties, and only a general correlation for the chromium content (Fig. 8.6).

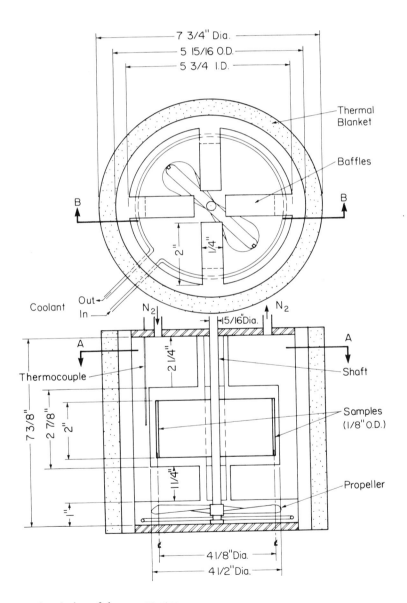

**Fig. 8.2** Cross section design of slurry pot tester

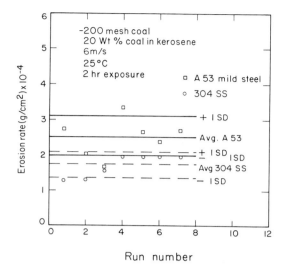

**Fig. 8.3** Reproducibility of erosion measurements in slurry pot tester

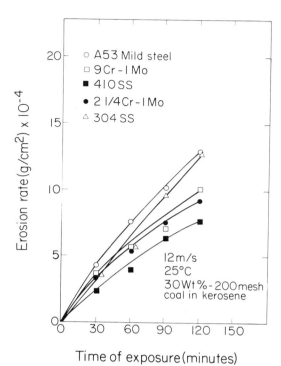

**Fig. 8.4** Cumulative erosion metal loss of several steel alloys at $V = 12$ m/s

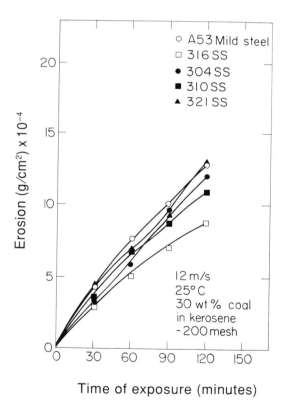

**Fig. 8.5** Cumulative erosion metal loss of several austenitic stainless steels and A53 mild steel at $V = 12$ m/s

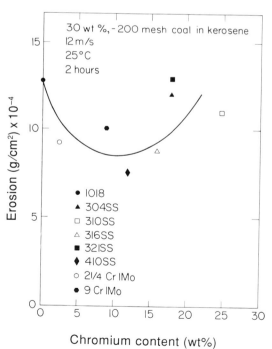

**Fig. 8.6** Correlation of erosion rate with chromium content of steels

**Effect of Particle Size and Solids Loading.** The effect of slurry velocity on the erosion of steel as a function of particle solids loading in the carrier liquid and particle size distribution are shown in Fig. 8.7 to 8.9. Figure 8.7 shows that both A53 mild steel and 304 stainless steel behave similarly in a fine, −200 mesh coal slurry. Erosion loss increases with slurry velocity. The particle size distributions of the fine and coarse coal particles were determined using a Tyler sieve analysis; they are listed in Table 8.2. The greater solids loading of the 50 wt% coal slurry results in a greater amount of erosion, even though its effective viscosity is higher than that of the 20 wt% coal slurry.

Figure 8.8 plots the amount of erosion when a coarse −30 mesh coal is used in the slurry. The larger size particles caused a considerably greater amount of erosion at the higher velocities than the small particles (Fig. 8.7). Also, the difference in erosivity between the two solids loadings is less for the larger size coal. The effect of velocity on the erosion is also markedly increased when the

coarser coal grind was used. Table 8.3 shows the velocity exponents for the two grinds of coal. The differences between the two coal grinds can be seen more graphically in Fig. 8.9 for the 20 wt% coal slurries. Four times as much erosion occurred on the samples exposed to the larger particles of coal than to the smaller particles at 12

**Table 8.2  Erodent particle size distribution, wt%**

| Size range, μm | −200 mesh | −30 mesh |
|---|---|---|
| >600 | 0.24 | 2.46 |
| 495-600 | 0.67 | 2.14 |
| 300-495 | 0.57 | 7.72 |
| 150-300 | 0.24 | 24.78 |
| 90-150 | 3.09 | 25.29 |
| 38-90 | 68.05 | 20.61 |
| ≤38 | 27.14 | 17.00 |

**Table 8.3  Erosion velocity exponents**

| Grind | Mesh size | A53 steel | 304 |
|---|---|---|---|
| 20 wt% coal-kerosene | −200 | 1.91 | 1.82 |
| | −30 | 2.97 | 2.83 |
| 50 wt% coal-kerosene | −200 | 1.68 | 2.15 |
| | −30 | 3.07 | 3.00 |

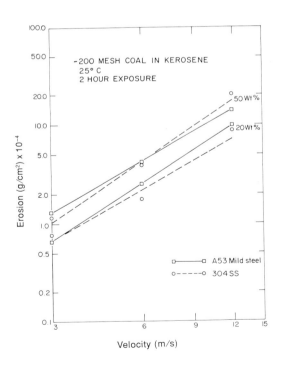

**Fig. 8.7**  Effect of velocity of two solids loadings of coal-kerosene slurries using fine, −200 mesh coal

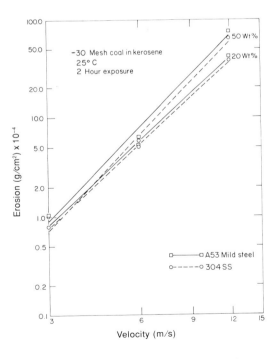

**Fig. 8.8**  Effect of velocity of two solids loadings of coal-kerosene slurries using coarse −30 mesh coal

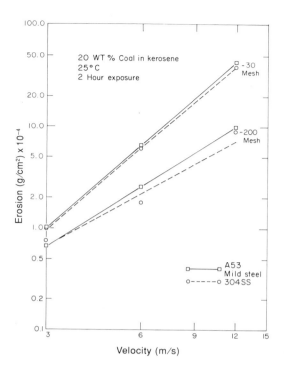

**Fig. 8.9** Comparison of coal particle sizes on erosion of steels in 20 wt% coal-kerosene slurry

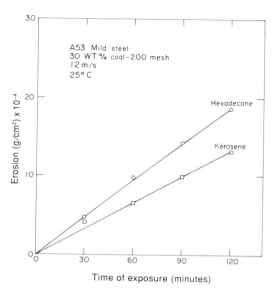

**Fig. 8.10** Erosion of A53 steel at 25 °C in coal-kerosene and coal-hexadecane slurries

m/s velocity (see Chapter 4 on the effect of particle size/kinetic energy). The difference between the two particle sizes is markedly less at the lower velocities.

**Effect of Boundary Lubrication of Liquid.** In order to determine the effect of the boundary lubrication of the carrier liquid on the erosivity of coal containing slurries, a solvent was selected whose boundary lubrication properties could be varied by means of a small addition that did not affect its viscosity appreciably. A long-chain hydrocarbon solvent, hexadecane, was used in these tests. The addition of 0.5 mol% of hexadecanoic acid to the solvent is known to increase the boundary lubrication of the solvent without changing its viscosity. The addition of the acid to the solvent polarizes the hydrocarbon chain, ordering the hydrocarbon chains perpendicular to the metal surface, imparting boundary lubrication. Shorter chain hydrocarbons and their acids cause a similar effect but at lower viscosities.

Figure 8.10 compares the erosion of A53 mild steel in a 30 wt%, −200 mesh coal-hexadecane slurry without acid addition to that of a 30 wt%, −200 mesh coal-kerosene slurry. It can be seen that the hexadecane-coal slurry is more erosive than the kerosene-coal slurry. Figure 8.11 shows the effect of the addition of the hexadecanoic acid addition to increase the boundary lubrication of the solvent in room-temperature slurry erosion tests of A53 and 304 steels. The increased boundary lubrication decreased the amount of erosion in a 120 min exposure by three times. Both alloys behaved in a similar manner. In a test series performed at 95 °C, the difference was more pronounced. Figure 8.12 shows that the erosion in the hexadecane plus acid slurry was one-fourth of that which occurred in the straight hexadecane for the A53 mild steel. A greater difference in erosion occurred between the two metals in the hexadecane slurry at 95 °C than occurred at room temperature.

**Effect of Temperature.** The elevated-temperature tests in a slurry pot tester for six steel alloys resulted in a behavior pattern that has also been observed at low elevated temperatures for

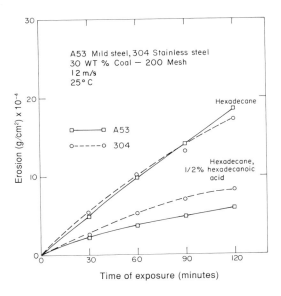

**Fig. 8.11** Erosion of A53 steel and 304 stainless steel at 25 °C in hexadecane and hexadecane + 0.5% hexadecanoic acid

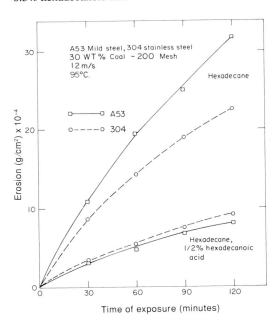

**Fig. 8.12** Erosion of A53 steel and 304 stainless steel at 95 °C in hexadecane and hexadecane + 0.5% hexadecanoic acid

ductile metals in gas-solid particle erosion tests (see Chapter 5). The erosion initially decreased with increasing test temperature for four of the alloys. Figure 8.13 shows the effect of tempera-

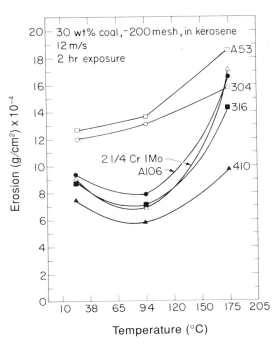

**Fig. 8.13** Erosion of steel alloys versus test temperature

ture on the alloys tested. The A53 and 304 alloys behaved in a similar manner, but their loss rates did not decrease in the 95 °C tests. Their rates of erosion were markedly higher than those of the other four alloys tested, particularly at room temperature and 95 °C. The A106 alloy is a finer grain size mild steel of the same nominal composition as the A53 pipe steel. Its lower erosion rate can be attributed to its finer grain size. The significant increase in erosion for all alloys at 175 °C is probably more the result of changes in the nature of the slurry than of changes in the erosion resistance of the alloys. A marked reduction in the viscosity of the liquid that generally occurs at elevated temperature could cause the increase to occur.

Figure 8.14 plots cumulative erosion versus time of exposure at 95 °C for a slurry velocity of 6 m/s for seven alloys. At the 5 h test time, the spread in the amount of erosion was significant; the highest eroding material had three times the erosion of the lowest. Figure 8.15 shows the cumulative erosion versus time curves for six of the alloys at a higher velocity (12 m/s at 95 °C).

It can be seen that the erosion markedly increases as the velocity of the slurry is increased, the A53

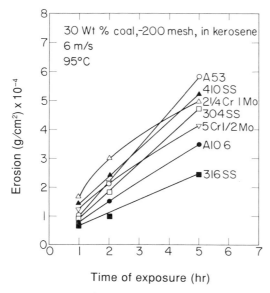

**Fig. 8.14** Cumulative erosion of steel alloys at 95 °C, $V = 6$ m/s

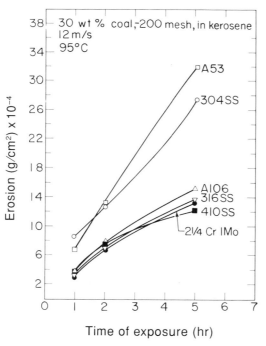

**Fig. 8.15** Cumulative erosion of steel alloys at 95 °C, $V = 12$ m/s

steel eroding five times more at 12 m/s than at 6 m/s after 5 h of exposure. The order of erosion rate changed among the alloys tested, and the A53 and 304 rates separated from those of the other four alloys, which were bunched together in a lower erosion range. When tested at 175 °C, the amount of erosion was up to eight times greater at 12 m/s velocity tests, compared to the 6 m/s velocity tests, as can be seen in Fig. 8.16.

Figures 8.14 to 8.16 are cumulative erosion rate curves for the various alloys tested. Several of the curves are straight lines, indicating that little comminution of the eroding particles occurred. However, some of the curves changed slope as the tests progressed. In order to gain more insight into the different shapes of the curves, incremental erosion rate curves were plotted using the same raw data. Figure 8.17 shows the incremental erosion rate curves for the alloys tested at 95 °C; the cumulative loss curves for these alloys are shown in Fig. 8.15.

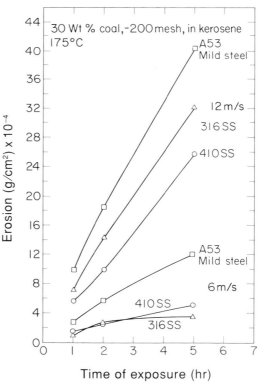

**Fig. 8.16** Comparison of erosion of steel alloys at 175 °C at two particle velocities

All of the alloys tested had their highest incremental erosion at the first weight measurement, after 1 h of testing. The rates for subsequent times decreased or leveled out toward a steady-state condition. The variations in shapes of the cumulative erosion curves corresponded with the variations in the shapes of the incremental curves. The straight line cumulative curve for A53 steel relates to the incremental curve for A53 (1018) that reached a steady state after 2 h of testing. The cumulative curve for 304 stainless steel that sloped up corresponds to the incremental curve that had a slight increase in rate at 5 h compared to its rate at 2 h. The other alloys that had the same concave downslope in their cumulative curves had decreasing incremental erosion throughout the 5 h test period.

**Metallography.** The texture of the eroded surface of a metal exposed to a liquid-solid particle eroding media and a gas-solid particle media are quite different. Figure 8.18 shows the surface of a 304 stainless steel specimen eroded by a 30 wt% coal-kerosene slurry. The surface is very smooth, almost polished, with few indentations

to indicate that 74 μm size particles had been striking it with sufficient force to remove material. In contrast, the surface of 1020 steel eroded by 200 μm SiC particles in an air-solid particle erosion test, shown in Fig. 8.19, is highly deformed (see Chapter 2). Platelets and shallow craters occur in profusion. The particle velocities were 15 m/s for the coal-kerosene slurry and 30 m/s for the SiC-air stream. While the differences in velocity and erosivity between the two erodent particles favor more severe erosion by the SiC, the striking difference between the surfaces shown in Figures 8.18 and 8.19 is still a valid comparison. The viscosity, boundary lubrication, and cooling capability of the kerosene compared to those of air are primarily responsible for the great difference in texture between the two surfaces.

A careful analysis of cross sections of the specimens eroded by the coal-kerosene slurry are required to observe the mechanism of erosion. Figure 8.20 shows a cross section of the surface of an A53 steel eroded specimen. It can be seen that shallow craters are present on the surface

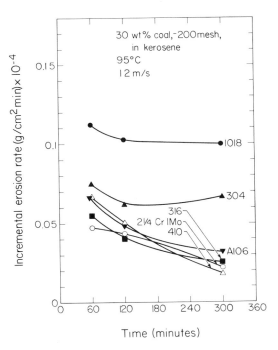

**Fig. 8.17** Incremental erosion of steel alloys at 95 °C, *V* = 12 m/s

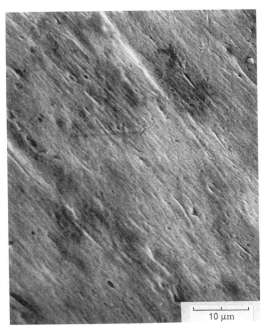

**Fig. 8.18** Eroded surface of 304 stainless steel from coal-kerosene slurry test

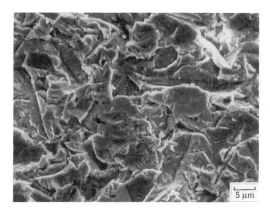

**Fig. 8.19** Surface of 1020 steel eroded in a SiC particle-air stream

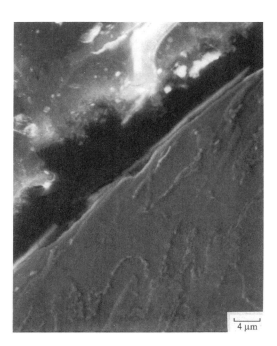

**Fig. 8.20** Cross section of A53 steel showing presence of platelets

along with their extruded platelets that lie parallel to the surface. In some places the platelets covered over nearby craters. Viewed from above, the surface at this cross section would appear similar to that in Fig. 8.18. Close inspection of the platelet at the left side of Fig. 8.20 shows part of it extending over the crater with a crack appearing in the platelet at the edge of the crater. The next particle to strike the area could knock off the part of the platelet extending over the crater. The long platelet on the right side of the photo shows that the platelets formed during slurry erosion have decidedly different geometries than those formed in gas-solid particle erosion. However, basically the erosion removal mechanism is the same in the two fluids.

### Discussion

**Effect of Alloy Composition.** The effect of alloy composition and standard mechanical properties on relative erosion behavior of the alloys tested was not indicated by the tests performed. The chromium content had some apparent relationship with the erosion rate as was shown in Fig. 8.6, but the effect was not a strong one. At 18% Cr content, the erosion rates varied more than 50% for the three alloys tested.

While the ductility of the alloys tested in gas-solid particle erosion (Ref 3) (see Chapter 3) can be used to some degree to relate the erosion behavior of different alloys with the more ductile alloys having lower erosion rates, no such correlation was possible in liquid-solid particle erosion behavior. The relatively low elongation of the A53 steel (25%), and the much higher elongation of the 304 stainless steel (55%) made no difference in their erosion rates, which were nearly the same in all of the room-temperature tests performed in coal-kerosene slurries.

Other properties of the alloys tested will have to be investigated further before correlations can be made that will be helpful to designers in alloy selection and materials developers to produce more erosion-resistant steels. For example, the strain-hardening coefficient of an alloy shows some promise of relating to erosion resistance.

**Effect of Particle Size and Solids Loading.** The effect of particle size and solids loading on the erosion rate of A53 steel and 304 stainless steel, the two highest erosion rate alloys tested, showed that more particles and larger particles generally resulted in more erosion (Fig. 8.7 to 8.9). However, the larger size particle (–30 mesh) eroded the steels more than the finer grind only at higher velocities (Fig. 8.9). At 3 m/s, both coal sizes caused nearly the same amount of erosion. Also, while the finer-sized particles showed a consistent difference in erosivity as a function of solids loading at all velocities (Fig. 8.7), the larger coal particles showed a difference only at the higher velocities (Fig. 8.8).

The combination of velocity and particle size effects on erosion relates to the kinetic energy of the particles and the degree of effective protection that the viscosity of the liquid provides to the metal surface from the particles trying to impact it. The fact that the higher effective viscosity of the total 50 wt% coal slurry caused more erosion to occur than the total 20 wt% coal slurry indicates that it is the viscosity of the liquid and not that of the total slurry that affects the erosivity of the slurry.

Apparently there is some critical level of particle kinetic energy at which the insulating capability of the liquid is reduced enough to permit a more direct relationship between the kinetic energy of the particles and erosion of the metals. At the lowest test velocity used, the mass difference between −30 mesh and −200 mesh size particles is not sufficient to change the kinetic energy of the particles enough to cause a significant difference in the erosion rates (Fig. 8.9). However at the higher velocities, the mass difference between the two sizes of particles is enough to affect their kinetic energy to the degree necessary to result in different erosion rates.

**Effect of Boundary Lubrication.** The ability of the carrier liquid to protect the impacting particles from the target metal surface was investigated by using a liquid whose boundary lubrication effect could be separated from its viscosity, which can also affect the ability of the particles to impart their kinetic energy to the target metal (Ref 4). The two curves, Fig. 8.11 and Fig. 8.12, which show the results of the experiments very effectively, indicate the role of the boundary lubrication of the liquid in modifying the erosivity of the slurry. The change to a greater level of boundary lubrication as a result of adding the small amount of acid lubricated the interface between the impacting particles and the target surface, significantly reducing their ability to indent the metal when they struck. Rather, they

had a greater tendency to only slide along the lubricated surface. This decreased the erosivity of the slurry to less than one-third of its level in the untreated solvent.

**Effect of Temperature.** The effect of temperature on the slurry erosion behavior of steels in the temperature regime used in this study (25 to 175 °C), follows a pattern first seen in gas-solid particle erosion (Ref 5) (see Chapter 5). Four of the six alloys plotted in Fig. 8.13 had lower erosion rates at slightly elevated temperatures than they had at room temperature. This effect could be due to interactions between the solvent and the coal particles that effectively increased the viscosity of the slurry, or to small but important modifications in behavior of the alloys in resisting erosion. It is not possible to know exactly what caused the decrease in the amount of erosion. In the case of gas-solid particle erosion, it is more likely that increased ductility of the alloys tested at elevated temperatures was responsible for decreased erosion.

The large spread in the amount of erosion occurring on the steel alloys tested that had similar mechanical properties (Fig. 8.14 to 8.16), as well as the changes in ranking of metal loss, cannot be readily explained. An important observation can be made based upon the incremental erosion curves in Fig. 8.17. The fact that the erosion decreased as the test proceeded indicates that there was a significant difference between at least part of the mechanism of erosion in liquid-solid particle erosion compared to gas-solid particle erosion.

In gas-solid particle erosion (Ref 6), the erosion process causes the surface to heat up as a result of the plastic deformation that occurs when the eroding particles strike the surface. Beneath the hot, eroding surface the energy of the particles is sufficient to cause subsurface cold working of the metal. The combination of the heated surface and the subsurface cold-worked zone results in a pattern of incremental erosion that is lowest at the beginning of the test and subsequently increases to a higher steady-state erosion (see Chapter 2).

Figure 8.17 shows that this pattern of erosion does not occur in liquid-solid particle erosion.

The highest incremental rate occurred near the beginning of the test and the erosion rate decreased subsequently. This behavior implies that the surface was being deformed into a smoother microstructure (see Fig. 8.18). As the test proceeded, there were fewer and fewer exposed platelets that could be struck and fractured by impacting particles (see Chapter 3). This retards the erosion process.

The reason for this difference between gas-solid particle erosion and liquid-solid particle erosion is thought to be the cooling effect of the liquid. Its much greater cooling capacity, compared to that of a gas, keeps the surface from being heated by the plastic deformation that occurs.

## Jet Impact Slurry Erosion of Steels

The slurry pot tester is a good screening device for determining the nature and relative behavior of materials exposed to solid particle containing liquid slurries impinging on their surfaces. However, a nearer simulation to in-service conditions that embodies more of the fluid mechanics aspects of slurry erosion is achieved when a defined jet of slurry is made to impinge on the test material surface.

### Experimental Conditions

In order to determine the effects of precise variations in slurry flow conditions on the erosion of metals, hard coatings, and ceramics, a test device was designed and constructed that directs measured quantities of slurry at flat specimen surfaces positioned at specific angles to the flow direction (Ref 7). In the jet impact tester (JIT) the exposure time can be varied as can the type of slurry and its impact conditions of velocity and angle. Quantities of slurry ranging from 1 to 80 U.S. gal containing selected solids loadings can be passed through the nozzle of the device to impinge on the specimen in a single test cycle at room temperature.

Figure 8.21 is a diagram of the JIT with the principal elements designated. The equipment operates by air or inert gas pressurizing of the stirred slurry hold tank, which forces the slurry through the nozzle (3 mm in diameter) into the test enclosure. The test tank contains a specimen holder that positions the specimen under the nozzle, approximately 1/2 in. below the exit of the nozzle. The holder can be positioned at any impact angle and holds flat specimens.

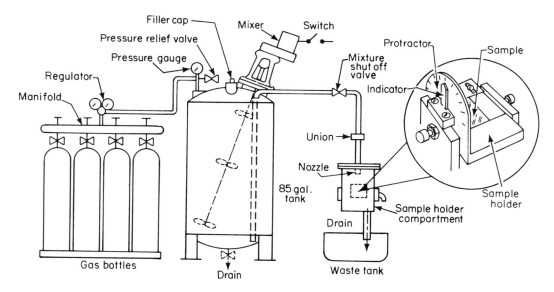

**Fig. 8.21**  Diagram of jet impact tester

The amount of slurry used in a single exposure is controlled by timing the release of the slurry whose velocity has been calibrated against the pressure level of the hold tank. The on-off valve in the nozzle assembly is used to control precisely the release of the slurry through the nozzle. The diameter of the nozzle was periodically checked with a plug gage after calibration to maintain the velocity-pressure relationship. The reproducibility of the test results from the JIT was determined in a series of five tests using a coal-kerosene slurry containing 30 wt% of –200 mesh coal to

erode 1020 steel at 13 m/s at an impact angle of 90°. Eighteen U.S. gal of slurry were used for each test in an 11 min exposure. Two of the specimens had a mass loss of 0.7 mg, and the other three specimens had a mass loss of 0.8 mg. On the basis of this level of reproducibility, single specimens were used for each test condition.

In order to determine the amount of eroding slurry to be used in each test, a series of incremental erosion tests were carried out. Figure 8.22 shows the results of the test. Steady-state erosion was achieved after 18 U.S. gal of slurry impinged on the surface. Therefore, 18 U.S. gal were used in all subsequent tests.

The alloys used in the test series are listed in Table 8.4. Test specimens 3.2 by 2.0 by 0.3 cm in size were prepared, and the surfaces to be eroded were polished with 400 grit SiC. After exposure to 18 U.S. gal of the coal-kerosene slurry containing 30 wt% of –200 mesh coal at 25 °C at selected velocities and impact angles, the specimens were removed from the test chamber, cleaned with soapy water and ethyl alcohol, and immediately weighed on a balance sensitive to 0.1 mg.

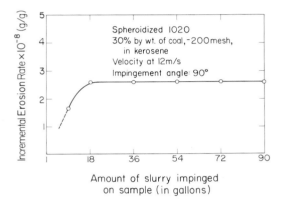

**Fig. 8.22** Incremental erosion test of 1020 steel

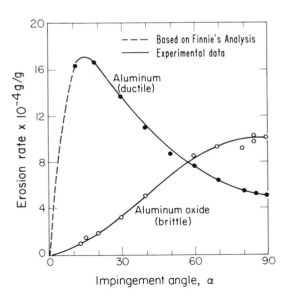

**Fig. 8.23** Effect of impact angle on the gas-solid particle erosion of ductile and brittle materials

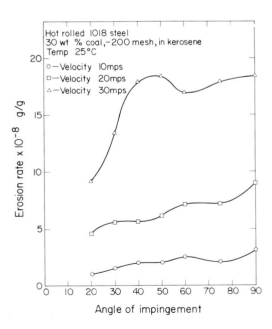

**Fig. 8.24** Effect of particle impact angle and velocity on slurry erosion rate of hot-rolled 1018 steel

**Table 8.4 Alloy composition and condition**

| Alloy | Nominal composition, wt% | | | | | | | | Condition |
| | Cr | Ni | Mo | Si | Mn | C | P/S max | Fe | |
|---|---|---|---|---|---|---|---|---|---|
| 1018-1020 | ... | ... | ... | ... | 0.5 | 0.2 | 0.09 | bal | Hot rolled, spheroidized |
| 4340 | 0.8 | 1.8 | 0.3 | 0.3 | 0.7 | 0.4 | 0.8 | bal | As-quenched, 200 °C temper, spheroidized |
| 2.25Cr-1Mo | 2.2 | ... | 0.9 | 0.3 | 0.4 | 0.2 | 0.02 | bal | Annealed |
| 5Cr-0.5Mo | 5.1 | ... | 0.6 | 0.02 | 0.5 | 0.1 | 0.02 | bal | Annealed |
| 410 SS | 12 | ... | ... | 1.0 | 1.0 | 0.2 | 0.1 | bal | Annealed |
| 304 SS | 18 | 9 | ... | 1.0 | 2.0 | 0.1 | 0.1 | bal | As-rolled |
| 321 SS | 18 | 10 | 0.4 (Ti) | 1.0 | 2.0 | 0.1 | 0.1 | bal | As-rolled |

## Impact Angle and Velocity Effects[Ref 1]

Determining the effect of impact angle on the erosion of ductile metals is generally indicative of the basic mechanism of erosion. Figure 8.23 shows the classical curve from Finnie's work (Ref 8) using SiC particles in an air stream for ductile and brittle materials. It can be seen that ductile materials reach a peak erosion rate around an impact angle of 20° with a decrease in erosion rate at higher impact angles. Brittle materials behave differently, reaching a peak erosion rate at a 90° angle.

Figure 8.24 shows the curve obtained for hot-rolled 1018 steel, a ductile metal, eroded by a coal-kerosene slurry at three velocities. The pattern of erosion versus impact angle is quite different from that shown in Fig. 8.23. The erosion rate increased with the impact angle, reaching a peak erosion rate at 90°. An intermediate peak of erosion occurred at an angle of 45° at the highest slurry velocity. This pattern of erosion in slurry flows was found to be consistent for all of the steels tested and has been observed by others (Ref 9). The marked difference in the effect of impact angle on the erosion of ductile metals between gas-solid particle and liquid-solid particle flows is directly related to the ability of the eroding particles to penetrate the lubricating film of the carrier liquid so that they can deform the target metal surface. This is most effectively done at the 90° impact angle. The mechanism will be discussed further later.

The effect of velocity on the erosion rates of 1018 steel specimens can be seen in Fig. 8.25. The velocity exponent, $n$, varied with the impact angle, generally decreasing with an increasing angle. The value of $n$ is less than that measured in gas-solid particle erosion tests of ductile metals (Ref 8). In other slurry erosion tests in a slurry pot tester (Ref 10), it was found that the velocity exponent was approximately 2.

The erosion rate curves for mild steel and low alloy steel AISI 4340 are compared in Fig. 8.26. Both steels were tested in the spheroidized condition. The 4340 steel had a tensile strength of 100 klbf/ in.$^2$ and an elongation of 25% in this condition, while the 1020 steel had a lower tensile strength of 57 klbf/in.$^2$ and an elongation of 36%. While the shapes of the two curves in Fig. 8.26 are somewhat different in the intermediate peak region, they both have approximately the same erosion behavior at the various impact angles. This occurs in spite of the marked differences in tensile strength and elongation. The stronger but less ductile 4340 steel has a higher erosion rate at the lower impact angles where the shear forces on the eroding surface are higher and about the same erosion rate as the 1020 steel at the higher impact angles. The peak erosion region of the 4340 steel occurs over a wider range of impact angles than that of any of the other alloys tested.

The erosion of 4340 steel in three different heat-treatment conditions was carried out over a range of impact angles at room temperature. The purpose was to determine the effect of hardness, strength, elongation, and microstructure on the erosion behavior of the steel. Figure 8.27 shows the erosion rate curves of the three heat-treated conditions plotted against the impact angle. As in earlier tests with other alloys, the general trend is for the steel to undergo increasing erosion rate with impact angle with the highest erosion rate occurring at a 90° angle and an intermediate erosion peak occurring in the 40 to 60° angle range.

The major erosion rate difference occurring between the as-quenched and the tempered and spheroidized heat treatments is unlike that which

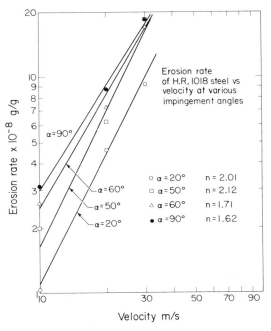

**Fig. 8.25** Effect of particle velocity on erosion rate of hot-rolled 1018 steel

occurs in gas-solid particle erosion of 4340 steel (Ref 3), where there was almost no effect of heat treatment on the erosion rate. The difference in elongation between the three heat-treated steels does not indicate that this property has a major effect on the erosion rate. If it did, the 200 °C temper curve would be much nearer the as-quenched curve. The Charpy impact strength of the as-quenched material and its fracture toughness are both significantly lower for the as-quenched material compared with those for the material tempered at 200 °C (Ref 7). While these properties did not affect gas-solid particle erosion rates, they do affect liquid-solid particle erosion rates. The relatively low strength of the spheroidized steel may account for its somewhat higher erosion rate compared with the 200 °C temper steel.

The effect of the impact angle on the erosion of 2.25Cr-1Mo alloy steel that is commonly used in chemical process plants is shown in Fig. 8.28. The erosion curve follows the pattern of increasing erosion rate with impact angle with an intermediate peak at approximately 45°. Unlike the 1018 steel curve in Fig. 8.24, the curves for both 10 and 20 m/s have an intermediate peak. The

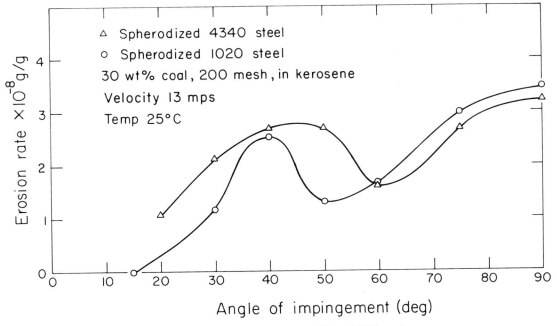

**Fig. 8.26** Effect of impact angle on erosion of spheroidized 1020 and 4340 steels

difference in the erosion rates between the two velocities is nearly constant over the range of impact angles. The erosion of 5Cr-0.5Mo steel follows the same pattern.

The erosion rates for the two chromium bearing steels are nearly the same at the lower velocity, but the erosion rates of the 5Cr-0.5Mo steel at 20 m/s are consistently lower than those of the 2.25Cr-1Mo steel. Both of the low chromium steels had about the same erosion rate as the 1020 steel at a velocity of 10 m/s but the 2.25Cr-1Mo steel had a higher rate at 20 m/s. The effect of the composition and morphology variations of the steels tested only affected their erosion behavior at the higher particle velocity.

Figure 8.29 shows the effect of impact angle on the erosion of 410 stainless steel at three different velocities. The curves have the same pattern as those for 1018 steel shown in Fig. 8.24. The intermediate peak at 30 m/s occurs at a 60° impact angle, which is considerably higher than the peak for the 1018 steel. The erosion rates of the 410 stainless steel at the slurry velocities used also differed from those of the 1018 steel, being lower at the lower two test velocities and nearly the same at the highest velocity. The rates were

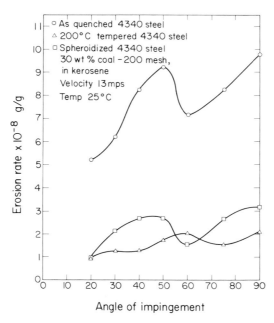

**Fig. 8.27** Effect of impact angle on 4340 steels at three heat treatments

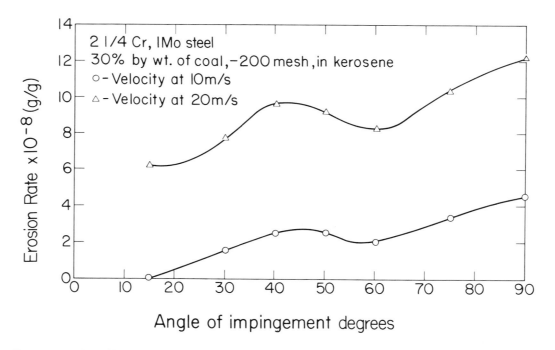

**Fig. 8.28** Effect of impact angle on erosion of 2.25Cr-1Mo steel

generally lower than those for the 2.25Cr-1Mo and 5Cr-0.5Mo steels.

Figure 8.30 shows the effect of impact angle on the erosion of two commonly used austenitic

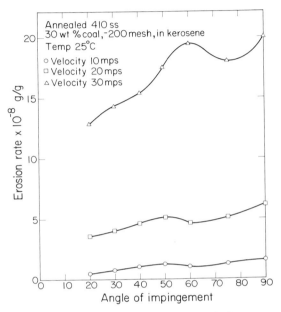

**Fig. 8.29** Effect of impact angle and velocity on erosion of 410 stainless steel

stainless steels. While the curves have the same general shape as those of the ferritic steels, they both exhibited a much more distinct intermediate peak, at a 45° angle. Their erosion rates were less than those of the ferritic steels, particularly at the intermediate peak impact angle and higher. The intermediate peak erosion rate was the same as the rate at the 90° angle for both steels.

### Slurry Characteristics Effects

In order to determine the effect of the viscosity of the liquid and the strength of the particles on the erosivity of the coal-kerosene slurry, a slurry was prepared of a lower viscosity liquid (water), and a higher strength particle ($SiO_2$). The $SiO_2$ particles had a finer particle size: 44 μm average diameter compared with 74 μm average diameter for the coal. The coal-kerosene and $SiO_2$-water slurries were used to erode 1018 plain carbon steel in the spheroidized and hot-rolled conditions, respectively. The difference in properties of the two conditions of the steel is not significant for the purpose of the tests. The resulting curves are shown in Fig. 8.31. (The break in the scale on the ordinate should be noted.) The $SiO_2$-water slurry caused more than 10 times the erosion of

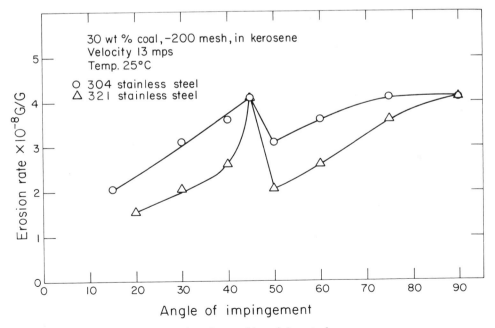

**Fig. 8.30** Effect of impact angle on erosion of austenitic stainless steels

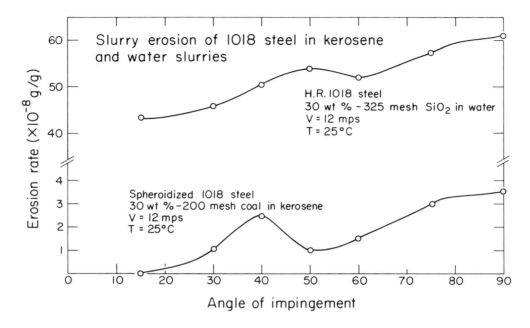

**Fig. 8.31** Effect of impact angle on the erosion of 1018 steel from coal-kerosene and SiO₂-water slurries

the higher viscosity, lower particle strength, coal-kerosene slurry. The pattern of the curve was the same as that from all other slurry test results with the intermediate erosion peak of the $SiO_2$-water slurry at 50° impact angle and the peak for the coal-kerosene slurry at 40°.

A further comparison was made of erosion rates using fluids of a still wider range of effective viscosity by testing 1018 mild steel specimens at the same flow conditions using coal-kerosene in one test and coal char-air in another test using the nozzle tester described in Chapter 2. Coal char is somewhat more erosive than coal; it was used instead of coal because the clumping tendency of the coal particles prevented them from flowing out of the hopper in the air-solid particle erosion tester. Both the coal and the char have considerably lower erosivity than particles such as sand and alumina. The particle sizes of the coal and the char were nearly the same, that is, approximately 100 μm in diameter. The tests were performed at a velocity of 30 m/s and an impact angle of 90° at 25 °C. The air-coal char erodent stream resulted in an erosion rate, $4.0 \times 10^{-6}$ g/g, an order of magnitude higher than the coal-kerosene erodent stream, which had an erosion rate of $2.0 \times 10^{-7}$ g/g.

### Microscopic Analysis

Microscopic examination of the eroded surfaces of ductile metals after exposure to small solid particles carried in either liquid or gas streams shows the same mechanism to have taken place. The eroded surface shown in Fig. 8.19 from a gas-solid particle test should be compared with that of a 1020 steel specimen eroded by a 30 wt% coal-kerosene slurry shown in Fig. 8.32. Figure 8.32(b) is at the same magnification as Fig. 8.19. The slurry was directed at the test surface from a nozzle at a velocity of 15 m/s and an angle of 30°. The liquid-solid particle eroded surface has undergone markedly less deformation than the gas-solid particle eroded surface. The few larger marks on the surface in Fig. 8.32 that show some platelet formation were probably made by large particles of mineral oxide that were not cleaned or screened out of the −200 mesh coal that was used to make the slurry.

The small impressions of unique appearance that occur in clusters on eroded 304 stainless steel (Fig. 8.33) are indicative of what part of the coal is erosive. Coal itself is too weak to erode steels at a measurable rate in the type of test used in this investigation. It is the mineral content of the coal

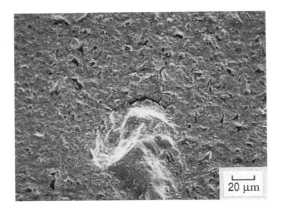

**(a)**

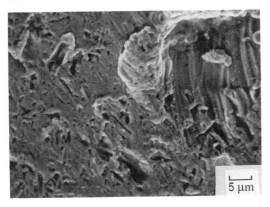

**(b)**

**Fig. 8.32** Eroded surface of 1018 steel from coal-kerosene slurry test at (a) lower magnification and at (b) higher magnification

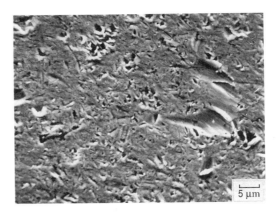

**Fig. 8.33** Eroded surface of 304 stainless steel from coal-kerosene slurry test

**Table 8.5   1018 Steel erosion rates in gas and liquid-particle streams**

| Particle | Fluid | Erosion rate, g/g | Test conditions |
|----------|-------|-------------------|-----------------|
| Coal | Kerosene | $2 \times 10^{-7}$ | $V = 30$ m/s for coal and char |
| Sand | Water | $6 \times 10^{-7}$ | $V = 12$ m/s for sand, $\alpha = 90$ |
| Coal char | Air | $4 \times 10^{-6}$ | 150 μm particles |

that has the particle integrity to deform a metal surface that it strikes. Figure 8.34 (Ref 11), shows that the small patterned impressions in the steel in Fig. 8.34(b) are very similar in shape and size to a grouping of framboidal pyrite clusters that occur in coal particles (Fig. 8.34a). Thus the coal acts somewhat as a tool holder to maintain the position of the framboidal erodent as it impacts the metal surface. This phenomenon is shown at several locations over the metal surface in Fig. 8.33.

## Comparison of Gas- and Liquid-Small Solid Particle Erosive Flows

The principal differences between the erosion of ductile metals in gas-solid particle streams and liquid-solid particle slurries are the rate of erosion and the effect of the impact angle of the particle on the target surface. The erosion rates of ductile metals for the same type and size of eroding particles at the same flow velocities are an order of magnitude higher for gas-solid particle streams than for liquid-solid particle slurries, as shown in Table 8.5.

The highest rate of erosion occurs at a shallow angle of impact for metals eroded by gas-solid particle streams while for liquid-solid particle streams the highest erosion rate occurs at an impact angle of 90° with a lower peak occurring in the 40 to 60° range for many, but not all, alloys. Both of these major differences may be considered as being due to the barrier nature of the liquid carriers, which impedes the impact of the solid particles on the target metal surfaces.

## Effect of Viscosity

The two properties of the liquid in a slurry that account for the barrier effect against erosion are their viscosity and boundary lubrication. The

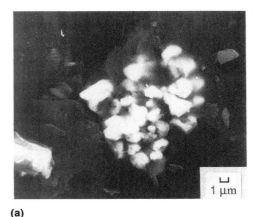

**(a)**

**(b)**

**Fig. 8.34** (a) Pyrite crystallites in coal. (b) Impression in eroded steel (1050 steel, $\alpha = 30°$, $V = 18$ m/s)

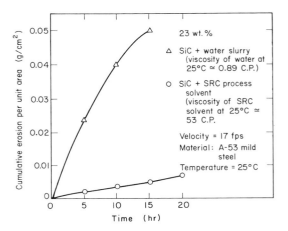

**Fig. 8.35** Effect of liquid viscosity on the erosion rate of A53 mild steel

physical concepts that account for the effects of carrier fluid viscosity on the erosivity of solid particle flows are known. The effect of viscosity is shown in Fig. 8.35 (Ref 4). It can be seen that the erosion rate was reduced by one order of magnitude by using the SRC-1 process solvent instead of water to erode A53 mild steel. The same order-of-magnitude difference was observed in the test results plotted in Fig. 8.31, although some of that difference was also due to the difference in the erosivity between the coal and sand particles used in those tests.

### *Effect of Boundary Lubrication*

How the effect of the boundary lubrication property of the liquid carrier fluid changes the shape of the erosion rate-impact angle curve com-

pared with the curve for gas-solid particle erosion is thought to be due to the following mechanism. At shallow impact angles, the force of the eroding particles is reduced by the lubricating film of the carrier fluid that is positioned between the particles and the eroding surface. As the impact angle increases, the particle momentum can more readily penetrate the barrier film of the liquid and the erosion rate increases. At an impact angle of 90°, the particles are least affected by the lubricating nature of the liquid and transmit a maximum impact force to the target surface. Thus, the highest erosion rate occurs at 90°.

The secondary erosion rate peak (lower than the maximum) that occurs on the steel alloys tested at impact angles between 40 and 60° can also be related to the lubricating qualities of the carrier fluid. In gas-solid particle erosion the amount of initial platelet extrusion and further extension by subsequent particle impacts are enhanced by the direction of the applied force of the particles (Ref 6). At shallow impact angles, between 20 and 30°, the effect of the vertical force component of the eroding particle that extrudes the metal beneath it is maximized by the horizontal component of the force that causes the extrusion to occur out from the point of impact, nearly parallel to the surface. This results in larger platelets which, in turn, cause greater erosion rates to occur at the peak erosion impact angles, as was discussed in Chapter 3.

In the liquid-solid particle slurries, the lubricating quality of the fluid decreases the eroding force at the shallower angles, around 20 to 30°. However, the basic effect, that is, the direction of the applied force enhancing the formation of larger platelets, still holds and a secondary peak still occurs but at an impact angle that is moved to higher angles, in the 40 to 60° range. Thus the occurrence of a peak erosion rate effect that is observed in gas-solid particle erosion at shallow angles also occurs in liquid-solid particle erosion, but at greater angles. The highest value of the erosion rate, however, is mitigated by the lubricating nature of the carrier fluid, and it is moved to the 90° impact angle where the erosive ability of the impacting particles is at its maximum. However, in one investigation (Ref 11) using coal-water slurries at the same slurry velocities, maximum erosion occurred at 60° impact angle. This was due to the lower viscosity and lubricating ability of the water carrier fluid compared to organic solvents.

Another aspect of the effect that the lubricating nature of the liquid carrier has on erosion by slurries is shown in Fig. 8.24. At the two lower velocities (10 and 20 m/s) the 1018 steel has essentially no intermediate peak. However, at the highest velocity used (30 m/s) an intermediate peak does occur. Apparently, at the lower velocities the impact force is low enough for the lubricating nature of the liquid to overcome the effect of the concentration of the impacting force and there is little, if any, enhancement of the formation of platelets. At the highest velocity, the force becomes great enough to affect the formation of the platelets. 410 stainless steel showed the same effect (see Fig. 8.29).

### Effects of Ductility

The effect of the ductility of the target steel on its erosion rate is shown in Fig. 8.26, in which 4340 low alloy steel and 1020 steel are compared. For impact angles up to 60°, the lower ductility but stronger 4340 steel has a somewhat higher erosion rate than the higher ductility 1020 steel. This effect of increasing ductility reducing the erosion rates of steels was discussed in Chapter 3 (Ref 3, 12). The same effect occurs in liquid-solid particle erosion, but to a lesser degree. At the

higher impact angles in the liquid slurries, where the effect of the lubricating aspect of the fluid is the least, the increased force of the particles on the target material appears to overcome the effect of the difference in ductility and the erosion rates of the two metals are essentially the same.

This effect relates to the remaining ductility in the plastically deformed platelets. At the lower impact angles in slurry erosion, more impacts on the formed platelets of the more ductile 1020 steel are required to fracture and remove them, resulting in a lower erosion rate. At the higher angles of impact, above 60°, the platelets formed in both the 1020 and the 4340 steel alloys are nearer their ultimate elongation on formation, and both alloys thereby produce platelets of the same vulnerability to being fractured and removed by subsequent impacts.

Another comparison between gas-solid particle and liquid-solid particle erosion is possible as the result of the data contained in Fig. 8.27. This figure shows that the low-ductility 4340, as-quenched steel with 8% elongation had a much higher rate of erosion than the steel in the more ductile 200 °C temper condition with 11% elongation and the still more ductile spheroidized annealed 4340 steel with 25% elongation. However, the higher ductility spheroidized steel tested had a higher rate of erosion than the less ductile 200 °C temper material. The attainment of a peak effectiveness of the ductility in reducing the erosion rate, with further increases in ductility at the expense of strength reversing the trend, was observed and is described in Chapter 3 (Ref 12) for gas-solid particle erosion.

### Erosion Mechanism

The microscopic analysis of the surfaces eroded by liquid-solid particle slurries indicated that the same basic mechanism of erosion occurs in both gas-solid and liquid-solid particle erosion of ductile metals. The formation, extension, and subsequent breaking-off of highly distressed platelets in gas-solid particle erosion are discussed extensively in Chapter 2 (Ref 6). The example of the same mechanism occurring in liquid-solid particle erosion shown in Fig. 8.20 is the logical extension of the gas-solid particle behavior.

Even though the erosion rate and impact angle effects between the two types of carrier fluids are markedly different, the same type of transfer of force from the impacting particle to the eroding ductile metal occurs, and thus the same mechanism of surface deformation and resulting erosion occurs. The effect of velocity on the erosion rate in slurry erosion also relates to the transfer of force from the impacting particle to the surface. The velocity exponent of 2 or less shown in Fig. 8.25 (Ref 1) for coal-kerosene slurries in the slurry pot type of erosion test relates to the kinetic energy $E_K = 1/2 \, mv^2$ of the impacting particles. The lubricating nature of the liquid medium changes the nature of the extrusion process, primarily by causing the platelet to extend more parallel to the metal surface and in a thinner cross section.

### Effect of Erodent

The carbonaceous component of coal is a relatively soft material with a hardness of 294 HV for bituminous coal. Because of this, its erosivity is relatively low. Chapter 4 (Ref 13) discusses the effect of particle strength on erosivity in gas-solid particle erosion. Comparing the strength of coal as an erodent with that of the weak erodent apatite (300 HV) indicates that the main bulk of the coal particle is not very erosive. However, coal contains many forms and sizes of much harder mineral matter, such as $SiO_2$, (700 to 1500 HV) and $Al_2O_3$ (1900 HV), which are very effective erodents. In the discussion of Fig. 8.32 to 8.34 the role of the hard mineral particles in the coal that are primarily responsible for the eroded material loss is presented. Improvements in coal-cleaning methods to remove the mineral ash for other reasons will also have a significant effect on reducing the erosivity of coal in slurry environments.

## References

1. Levy, A.V.; Jee, N.; and Yau, P., Erosion of Steels in Coal-Solvent Slurries, *Wear*, Vol 117 (No. 2), 1987, p 115-128

2. Levy, A.V., and Hickey, G., Liquid-Solid Particle Erosion of Steels, *Wear*, Vol 117 (No. 2), 1987, p 129-146

3. Foley, T. and Levy, A.V., The Erosion of Heat Treated Steels, *Wear*, Vol 91 (No. 1), 1983, p 45-64

4. Levy, A.V., Erosion-Corrosion of Metals in Coal Liquefaction Environments, *Mater. Perf.*, Vol 19 (No. 11), NACE, Nov 1980, p 45-51

5. Doyle, P.A., and Levy, A.V., The Elevated Temperature Erosion of 1100 Aluminum by Gas-Particle Stream, *Proc. Conf. Corrosion-Erosion Behavior of Materials* (St. Louis), Oct 1978, AIME-TMS, p 162-176

6. Levy, A.V., The Erosion of Metal Alloys and Their Scales, *Proc. NACE Conf. Corrosion-Erosion-Wear of Materials in Emerging Fossil Energy Systems* (Berkeley, CA), 1982, p 298-376

7. Li, S.K.; Humphrey, J.A.C.; and Levy, A.V., Erosive Wear of Ductile Metals by a Particle-Laden High Velocity Liquid Jet, *Wear*, Vol 73 (No. 2), 1981, p 295-309

8. Finnie, I., Erosion of Surfaces by Solid Particles, *Wear*, Vol 3, 1960, p 87-103

9. de Bree, S.E.M.; Begelinger, A.; and de Gee, A.W.J., "A Study of the Wear Behavior of Materials for Dredge Parts in Water-Sand Mixture," Paper E1, Proc. 3rd Int. Symp. on Dredging Technology, BHRA Fluid Engineering, Cranfield, Beds., 1980

10. Levy, A.V., and Hickey, G., "Surface Degradation of Metals in Simulated Synthetic Fuels Plant Environments," Paper 154, National Association of Corrosion Engineers Corrosion Conf. (Houston) May 1982

11. Sargent, G.; Spencer, D.; and Seques, A., Slurry Erosion of Materials, *Proc, NACE Conf. on Corrosion-Erosion-Wear of Materials in Emerging Energy Systems* (Berkeley, CA), Jan 1982, National Association of Corrosion Engineers, p 196-231

12. Levy, A.V., The Solid Particle Erosion Behavior of Steel as a Function of Microstructure, *Wear*, Vol 68 (No. 3), 1981, p 269-288

13. Levy, A.V., and Chik, P., The Effects of Erodent Composition and Shape on the Erosion of Steel, *Wear*, Vol 89, 1983, p 151-162

# INDEX